Shadworth Hollway Hodgson

The Philosophy of Reflection

Vol. 1

Shadworth Hollway Hodgson

The Philosophy of Reflection
Vol. 1

ISBN/EAN: 9783337236441

Printed in Europe, USA, Canada, Australia, Japan

Cover: Foto ©berggeist007 / pixelio.de

More available books at **www.hansebooks.com**

THE

PHILOSOPHY OF REFLECTION

BY

SHADWORTH H. HODGSON,

Hon. LL.D. Edin.

AUTHOR OF "TIME AND SPACE," "THE THEORY OF PRACTICE," ETC.

IN THREE BOOKS.

VOL. I.

CONTAINING BOOKS I. AND II.

LONDON:
LONGMANS, GREEN, AND CO.
1878.

TO

Samuel Taylor Coleridge,

MY FATHER IN PHILOSOPHY,

NOT SEEN BUT BELOVED.

CONTENTS OF VOL. I.

— ⚬ —

BOOK I.

ANALYSIS OF ASPECTS.

CHAPTER I.

PHILOSOPHY AND SCIENCE.

CHAPTER II.

PRIMARY, REFLECTIVE, AND DIRECT CONSCIOUSNESS.

CHAPTER III.

THINGS-IN-THEMSELVES AND PHENOMENA.

BOOK II.

ANALYSIS OF ELEMENTS.

CHAPTER IV.

PRESENTATION AND REPRESENTATION.

CHAPTER V.

PERCEPT AND CONCEPT.

CHAPTER VI.

CONTRADICTION AND CONTRARIETY.

PREFACE.

" Car notre siècle n'a pas encore tout dit."
Viollet-le-Duc.

THE purpose of these volumes is, first, to lay down the outlines, principles, and method of a system of Metaphysic, basing it upon known facts of consciousness; next, to show that this system necessitates the conception of a Constructive Branch of philosophy, dealing with phenomena which are but very partially accessible to us; and lastly, to combine these two branches (the latter given in the merest outline) into a single System of Philosophy.

As the present work completes for me a certain cycle of thought, I may, perhaps, be permitted to prefix some remarks on the scope and sources of my philosophy, which I abstained from doing when I published its predecessors, "Time and Space" (1865), and "The Theory of Practice" (1870). It seemed best to let those works speak for themselves. But now, as I gather up their results into a more systematic shape, a word or two of introduction may not be out of place. There is this advantage in a preface, that what is there said may often arrest attention, when the very same thing in the text, though possibly said just as well, may be passed over as unimportant or unintelligible, unless it happens at once to suggest to the reader some pre-conception of his own.

My present work, then, is the harvest of my two former ones; they are the standing corn, this is the stack. There is a pightel or two added besides. But it is far from being a mere abridgement or summing up of their results. A great deal is entirely omitted; all in fact that was not requisite to the logical method, articulation, and construction, of a system of philosophy. Ethical questions, for instance, and the analysis of emotion and character are hardly touched on. Some things are completely new, for instance, the proof of the axiom of uniformity, and the addition of the constructive branch of philosophy; but the fundamental conceptions, which are the basis of the whole, are clearly enunciated, and to a great extent built upon, in the two former works. The present work exhibits a system which, in gathering up and completing their results, at once shows to what philosophy they lead as their outcome, and is a means of verifying the truth of their facts and the justice of their conclusions. If *they* are true, *this* will be the consequence. Does it, or does it not, harmonise with the results of science, in branches other than metaphysic?

I deliberately maintain that it does. And then I say,—If so, we are at last in possession of a metaphysical system which will not have to be reversed, however much it may in the future be enlarged and differentiated. A secure basis, or nucleus, of philosophical doctrine has been found, which must be included in whatever future systems the course of time may bring with it. Metaphysic has been organised, and, within the limits drawn by metaphysic itself, philosophy has been constituted.

On this subject it would be unbecoming to speak without extreme caution, and even diffidence. It is a subject in which experience shows that men are peculiarly liable to self-deception. The numerous systems proposed with unhesitating confidence as the final solution of metaphysical problems, and all in turn abandoned or forgotten, naturally lead us to regard all similar attempts as fruits of hallucination, like the hallucinations which mathematicians ascribe to squarers of the circle and discoverers of perpetual motion. The wide-spread prejudice against metaphysic is therefore not without foundation; and it becomes necessary for any one who brings forward a new system in that matter carefully to ask himself whether he may not be the victim of self-deception, of an hallucination spreading not over portions only but over the whole subject, and making him see metaphysic as a branch of genuine knowledge, when in truth it is only a mirage of the imagination.

If, however, after such an enquiry he still sees reason for concluding that the *foundations* of metaphysic are certainly true, and that the nature of the subject is such as to explain the existence of hallucinations concerning it, on the supposition that they are so, then, I think, it is his duty as an enquirer to avow his conviction, and once more to come forward with a system which may be the means of bringing that conviction to the test. This, then, is my position. In spite of those repeated failures I still come forward with another metaphysical system.

The distrust of metaphysic, which I have already said has some foundation, though not a sufficient one, is accompanied with positive dislike. The very name is unpopular. No one who wished to make his way in the world would adopt it. Barring theologians who look upon it as a possible ally, or even entertain the hope of one day reducing it to the ancient position of handmaid to theology, there are few English writers of weight who, if they have occasion to allude to it, do not heap scorn and contumely on what they are pleased to call "metaphysics." There is, nevertheless, a considerable body of facts bound together by a method, which can bear no other name so properly as *Metaphysic* or *Metaphysical Philosophy*, meaning that analytic branch of knowledge to which Physic leads, and which in order of study comes after physical knowledge. The order of study coincides with the order which Physic and Metaphysic held to each other in the arrangement of Aristotle's writings, to which arrangement the name Metaphysic, Τὰ μετὰ τὰ φυσικὰ, owes its origin.

At the first, no doubt, and in Aristotle's own mind, metaphysic was identical with ontology. It was the study which investigated "the existent in respect of its existence, and its essential attributes in that respect," τὸ ὂν ᾗ ὂν καὶ τὰ τούτῳ ὑπάρχοντα καθ' αὐτό.[1] But when that study was stripped of its *absolute* character, when the "existent," which is its object, was shown to be relative and phenomenal, then the only proper name for it was Metaphysic, because *Ontology* carried with it a suggestion of absolute existence. Metaphysic is a general name for

---

[1] Ari t. Metaph. Γ. i. p. 1003, a. 21.

the study of existence which implies no assumptions or foregone conclusions.

Still the dislike to it, name and thing, will never be removed by remarks of this kind; and uphill work it must inevitably be, to win even a hearing for anything that shows metaphysical colours, when so many are committed to an attitude of contemptuous incredulity. Theology has a long train of clients; metaphysic few or none. Metaphysical writers who are not also theologians, on some traditional basis or other, are in England like them that cry in the wilderness, their voices once heard and then for ever inaudible. Such was the subtil and acute author of Thorndale, William Smith. Such bids fair to be the philosophic Grote, author of Exploratio Philosophica, notwithstanding that he has found an editor who from time to time publishes portions from his manuscripts. Many such, I do not doubt, have spoken and passed away. And now comes yet another, to raise their cry once more, once more perhaps to hear it die away unanswered. But the time will come, at last it will come, when a metaphysical system will make its truth acknowledged, and honour will be done to those who, in the midst of empiricism, have patiently though ineffectually reasoned on its principles, then at last recognised as true.

But to turn to the present system. The principle which I think I have established beyond the possibility of reversal is that of Reflection. Whatever other parts of the system may be found untenable, this will stand. It is nothing else than Descartes' *Cogito ergo sum*, analysed and applied. Also, it is this which makes the system a Metaphysic, and distinguishes it from an " *Erkenntnisstheorie*," or logic of cognition, which is a descendant of Kant's *Critic*, a preliminary of metaphysic. Reflection is the foundation of metaphysic because, being the moment of distinguishing the objective and subjective aspects of phenomena, it gives us our notion of *existence* as well as cognition, and that in the largest sense of the term existence, so that we cannot speak or even frame a notion of anything beyond it. Whatever notion we frame lies within it, is subjective as well as objective or imaginary. This notion of subjectivity was what I attempted to explain in Chapter I., and to enforce in the Epilogue, of "Time and Space."

Reflection, then, is the cardinal point of the system. Next I would call attention to the circumstance, that the relation of philosophy to science is shown in it to be a relation not of rivalry but of division of labour, founded on a philosophical principle of division, the distinction between *Nature* and *History*. For the first time, as I believe, a firm line has been drawn between science and philosophy, without sacrificing the necessary independence of either. If any choose, after this, to keep within the fruitful limits of pure science, it is quite open to them to do so,— πλατεῖα κέλευθος. But they are thereby setting themselves a voluntary limitation, which it is more catholic to overstep.

In close connection with this, I would next point to what I have already mentioned as a new feature of the present work, the Constructive Branch of philosophy. The old ontological questions are shown to belong to this branch, but only when stripped of their ontological or absolute character. A transformation of these problems has taken place, and it has been shown why they are not soluble in their old shape, and also in what shape they are conceivably soluble. The establishment of the Constructive Branch of philosophy, in dependence on Metaphysic, is the fulfilment of an anticipatory conception in "Time and Space,"[1] where I have represented metaphysic, not as the answer, but as the condition of prosecuting the enquiry into all such great philosophical questions as: Whence man and the world come, and whither they go, and what the meaning is of the whole scene of existence, as it unfolds itself before us.

It will be said, that after all I have not given the principle or source of existence of the actually known world. True. But why have I not? Because the Unseen World must be known first, before we can point to any one or more principles which *make* the existence of our seen world what it is. What Philosophy *can* do, I have (to the measure of my ability) done; with what success it is not for me to judge. I have not attempted what it cannot do. But I have shown both what things it can, and what it cannot do, and why.

Next I would remark, that all philosophical systems, standing as they do in various relations, present various distinct sides or facets to observation. They are not like pictures which show a surface only, but like sculpture which invites inspection from every side. I will name some of these facets, and the principal feature in each, presented by my own system. They are:

1. In relation to Science,—The distinction between Nature and History.

[1] Part I. Chapter I. p. 14.

2. In relation to Hume and Kant,—The principle of Distinction of Inseparables.
3. In relation to Hegel (the last great system
   not recognising a "Thing-in-itself"),—
   The nexus of Percepts as prior to the
   nexus of Concepts.
4. In relation to Ontology or Absolutism,—The
   distinction of a Constructive Branch,
   apart from the analytic, in philosophy.
5. In relation to the History of Philosophy at
   large (which is tantamount to taking the
   system *per se*, as a philosophical system
   simply).—The principle of Reflection.
6. In respect of the Logical Movement of
   thought within the system itself, and for
   avoiding fallacies,—The distinction between First and Second Intentions.

Some few additional remarks on the foregoing
list. First as to the method itself. This being analytic, its life and soul must consist in drawing distinctions. But to draw a distinction is to determine
a relation. The distinctions are therefore also bonds
of union between the things which they distinguish
from each other; and this is why distinguishing presents us with a *system*. The system is built up of
distinctions, and I have marked this peculiarity
by taking distinctions as the titles of the several
chapters.

In the logical movement of thought, the first and
most important distinction is that between First and
Second Intentions. The nature and import of this
may be shown by considering, in the case of words,
the two ways in which we use words significantly.

We use words either denotatively or connotatively; denotatively when a word is used as a mere mark or sign to point out *which* thing of all possible things we mean to speak of, and connotatively when it is used to point out a supposed characteristic of the thing denoted, so that, if that characteristic is *not* found in the thing denoted, then the connotation is false. For instance, I see an object in the road before me, and denote it by *that object yonder*. Going on farther, I connote it by that *milestone* yonder. But on coming nearer it turns out to be a man. Then the connotation *milestone* is false, and is shown to be so by holding fast the denotation.

Now to connect this with first and second intentions. Objects are expressed by words used denotatively, *that object yonder*, in order to bring them before us in their relation to consciousness alone, separately from other objects in consciousness. To use terms denotatively is simply to *designate* which thing you mean to look at, and to exclude all assumptions about it. True, you cannot do this without reference to the other, the excluded, objects; but you use the reference for the purpose of excluding the objects referred to. You then proceed to look at the object, so designated, *in its first intention*. It is an object in relation to consciousness alone.

But it does not follow that, when you perceive its features, and describe it by a word with a connotation, that then this word describes it in one of its second intentions. No. If the distinction between first and second intentions *coincided* with that between denotation and connotation, the former distinction would be superfluous. True, a connotative word implies a reference to other objects in consciousness;

but this is precisely parallel to the similar reference, just noticed, in the case of denotative words; we must abstract from this kind of reference to other and excluded objects, as a common necessity of language.

When the object which has been isolated by designation is distinctly perceived and described by a word with a connotation, the question arises,—does the connotative word describing it express a feature or features of the object as it is perceived, or does it express a relation or relations to other objects excluded from it by the original designation? If the first, the connotative word expressing it is a term of the first intention; if the latter, it is a term of second intention. Take, for instance, the milestone. I *see* that it is *stone*. If by this I mean to infer that it has the qualities of stone, the term *stone* is used of it in its second intention. If I mean only the perceived colour, hardness, and roughness, of the object before me, then the term *stone* is used of it in its first intention. It describes what I perceive and not what I infer. But there is no mark in the word *stone* to indicate in which of these uses it is employed. Hence the necessity of the distinction between first and second intentions. For if it had been already embodied in language, it would not need to be drawn and insisted on in philosophy.

The only so called *facts* of consciousness which can be appealed to as evidence are facts taken in their first intention. But very often facts of consciousness are appealed to as evidence, without this carefulness; some inference or hypothesis is included with them. For instance, the sense of freedom in choosing and acting. Taken strictly in its first intention. the words being used denotatively, the sense of freedom is a fact of consciousness. There is a familiar feeling or perception which is fairly called a sense of freedom. But if we have previously defined, or unconsciously assumed, freedom to mean action without motive, or a particular kind of causality opposed to physical causality, then the sense of freedom is no longer a fact of consciousness of the kind which can be appealed to in evidence, and can no longer be drawn in to show the existence of an immaterial Self, or a faculty of Will. The sense of freedom in its first intention does not include the knowledge that it is a Self which is free, or that there is a Will which exerts the power of free choice.

Closely connected with the foregoing is the method of answering objections and detecting fallacies, which may be called the *Point of View* doctrine. The structure and mechanism of consciousness is such, that we are constantly adopting a particular point of view without being aware of what we are doing. I mean that we are always unconsciously making some assumption, or bringing the object of discourse into connection with something, which there is nothing to justify us in doing. Hume, for instance, was perfectly unconscious that he was making an assumption when he spoke of impressions as separate, as if they came raining in upon us like hailstones. You can see it was unconscious, just because he does not *say* that they come separately.

The commonest form of fallacy perhaps is to adopt some well known and legitimate point of view, and forget its connections and limitations. Thus many a scientist takes his stand on the historical method, in itself so legitimate and so useful, and is unable to see

that it has pre-suppositions which condition it, and
which, even supposing it to be the final method of
science, unavoidably subordinate it to a larger enquiry
which is philosophy.

Another instance is the assertion of contingency
in phenomena and events. This assertion arises from
adopting a point of view looking forward into the
future, and assuming that this view of things should
control views taken from other points; whereas it
is equally essential to thought to look at the course
of time statically, transversely as it were, whereby what appeared to be a process is frozen into
fixity, arrested into an eternally enduring present
time.

To say that phenomena change, and give place to
different phenomena, is only another way of saying
that our point of view in looking at the same phenomena changes. To enumerate and systematise the
principal possible points of view is to enumerate and
systematise the corresponding sides presented by
phenomena. The question is, what points of view
are partial and permissible, what unavoidable and
obligatory; what are dominant and what conditioned;
what the including and what the included.

Here it is that Reflection vindicates its place.
There is no more obligatory, dominant, and comprehensive point of view in all the circuit of knowledge
than that of reflection. We adopt it when we put
our first question to an object in its first intention.
And there is no result, either of science or of philosophy itself, to which the question *What?* may not
be put by reflection, no result which is not in its
turn the starting-point of new enquiries. To bring
this process out in its necessity, to analyse its mechan-

ism, to show some of its consequences,—this is the business of the present book.

These remarks will not, I am aware, be fully intelligible before the book has been read; but they will serve, I hope, to guide a judgment upon it after a first survey. It is not my intention to enumerate beforehand all its salient points, or sum up what I conceive to be its results. In what remains of the Preface, I shall content myself with trying to make my readers acquainted with my mode of thinking, and put them as it were on my track.

In philosophy we must be prepared for surprises. Most admirable is the genial nonchalance of Hume, when he confesses the utter failure of his system to explain anything, its total break up into inevitable contradictions. Good luck to those who never trouble their heads with philosophy; but for himself, he shall continue to philosophise. This scepticism? It is scepticism defied. Could anything show a more profound, a more absolute, faith in reason, a more utter confidence that reason was incapable of deceit? It is *Le Roi est mort*, instantly followed by *Vive le Roi*. This is Hume's tone in the *Conclusion* of the First Book of the Treatise of Human Nature; and it is in perfect harmony with the *Introduction*, " And indeed nothing but the most determined scepticism, along with a great degree of indolence, can justify this aversion to metaphysics." Be it known, then, that there are two sorts of scepticism, one, the philosophical, *a posteriori*, sort, which is the result of reasoning badly, and is based upon a profound trust in reason; the other, the *a priori*, philistine sort, which springs from a distrust of reason, and is compensated by an equivalent confidence in creeds.

Observe, I speak of the Hume of the Treatise, the work of his early manhood, published in 1739. Nine years later, in 1748, when the Enquiry concerning Human Understanding, which is the First Book of the Treatise recast, was published, Hume was a different man. He had suffered the unjust neglect of the world to spoil him. He was piqued, and drew back into his shell. His philosophy is still the same, both in its principles and its results; but it is now presented in a manner to secure its results against antagonists, without the frank disclosure of its methods, the *arcana imperii.* In fact, on the one hand he was committed to the results, and on the other, the public had refused to listen to the methods. In the Treatise he shows his mind at work, in the Enquiry he defies you to get within his guard.

True, to estimate Hume fairly as an individual, to compare him fairly with other philosophers, every stage of his philosophical development, the intellectual power displayed in the last as well as in the first, what he retracted as well as what he stood by, must be taken into account. That will give us the Hume of biography, the Hume of literary history. But the Hume that belongs to the history of philosophy, the Hume that roused Kant from his "dogmatic slumber," will always be best known to us from the Treatise of Human Nature. The doubts and the reasonings which led to them, once made public, could no longer be withdrawn. They were there, working in the minds of others, totally irrespective of the future position of their author to friends or foes.

But to go back to the two sorts of scepticism.

Hume's was of the philosophical, *a posteriori*, sort.
He could therefore sit loose to the fate of his own
speculations, secure that, if they were not wholly
true, they could give place only to a greater truth.
A Baconian *Sic cogitavit* guided his pen; hence came
that freedom in making admissions, that playing with
words like two-edged blades and no handles, that
delicate irony, which made Kant call him " the subtil
and gentle Hume," *der feine und sanfte Hume*, and
which others have misinterpreted as signs, if not of
insincerity, at least of levity unbecoming a philosopher.[1] He saw the contradictory side of things,
inconsistencies were everywhere before him. You
can almost hear him confessing with Swift, "Mihi
ludibria rerum mortalium cunctis in negotiis obser-
vantur." I say that Hume had this temper, which
was not unphilosophical, *because* he was firmly
grounded in the assurance that reason must rule, in
other words, because he had *faith in reason*.

Having this faith, a man may sit loose to the fate
of any philosophical system or doctrine, even those
for which he is individually the most concerned.
One thing only is imperatively forbidden—the *a
priori* scepticism. One foregone conclusion is imperatively necessary—to have no foregone conclusions.
Have faith in reason.

I have often thought what a subject it would be
for the genius of a Bewick or a Doré, a tournament
with philosophical systems, figured as extinguishers;
the knights laying their extinguishers in rest, and
tilting against each other with the large and open

[1] For instance, my friend Mr. Collyns Simon. See his article
La Religione e la Metafisica in *Filosofia della Scuole Italiane*,
Vol. XV. Disp. I. 1877.

ends, till victory declared itself for the largest-
rimmed extinguisher that was not split up or shattered
in the encounter. How Leibniz' would disappear
into Kant's, how Comte's would be engulphed in the
depths of Hegel's. How merrily should the combat-
ants feast together afterwards, and what trumpets
might be made of the extinguishers; what an organ
they would compose; what music of the spheres
would pour from their united throats.

But at this point I think I hear a reader say to
me, Tell us plainly what school you belong to. Hegel
we know, and Hume we know, and even of Kant we
have a sort of inkling; but your opinions seem to be
a farrago which can be held by no sane man.

Courteous reader. I belong to a school which has
set itself to carry farther the *critical* strain in Kant's
speculations, just as Hume carried farther Berkeley's
method without stopping at Berkeley's results. The
master of this school is Salomon Maimon, and in that
school I am a disciple. Not indeed a disciple of
Maimon's from the first, for I entered without knowing it upon the track, which he was the first to
strike into more than three quarters of a century ago.
I first became acquainted with Maimon's writings
through Herr Kuno Fischer's History of Modern Philosophy.[1] And I made no systematic study of them
until the greater part of the present work had been
actually written. Then I saw in how many and in
how important points I had been anticipated by Maimon; he writing as a contemporary of Kant, I after
enjoying the instruction of the whole post-Kantian
philosophy down to Hegel and Schopenhauer.

Maimon continued the Kantian criticism, and pur-

[1] Vol. V. Erste Abtheilung, published in 1869.

sued it to new results, namely, to a system founded on his principle of Determinability, *Bestimmbarkeit.* Kant had introduced *Critic,* name and thing; it was a branch of analysis, like *Logic,* but having for its special purpose to determine the adequacy of Reason to its problems, its power to perform what it spontaneously undertook; and not merely, like Logic, to determine its structure and method.

The *Critic* of Pure Reason is an analytic enquiry into the power and capacity of Reason. Maimon continued this enquiry. Whereas Fichte, Hegel after him, and Schopenhauer, took up Kant's *results,* and constructed a theory of Ontology on them and out of them.

An age which ran after Hegel could not have much attention to bestow upon Maimon. Still it is probable that so subtil and so acute a mind as Maimon's, placed moreover at so central a point of view, was not without effect on his age, though not one that came to the surface. As he says himself: "A writer who has a good *style* is read. One who has *expository* power is studied. One who has neither the one nor the other, supposing him in possession of weighty and new truths, is used. His *mind*, though not his *name*, is imperishable."[1]

Thy *name* too, Maimon, if any words of mine could celebrate it. But he who now writes has a pen as little potent as thine own.

---

[1] Philosophisches Wörterbuch, 1791 : sub voc. *Vortrag und Stil*, p. 155. "Ein Schriftsteller der einen guten *Stil* hat, wird gelesen. Der einen guten *Vortrag* hat, wird studirt. Der beide nicht hat, wird, wenn er nur im Besitze wichtiger und neuer Wahrheiten ist, benutzt. Sein *Geist*, nicht aber sein *Nahme*, ist unsterblich."

My philosophy then belongs to the Critical School of Maimon, as the continuer of Hume and Kant, so far as its structure and method are concerned. But there is another source to be mentioned, to which it owes, if I may so speak, its informing spirit, and the suggestion of some of its most distinctive conceptions. My relation to Coleridge is very different from my relation to Maimon. From Maimon I learnt (as it happened) nothing; from Coleridge everything, for he taught me what things were worth knowing, and where they could be found. It was in very early life that I met with his writings, and they became to me almost as oracles of God. Chiefly the Aids to Reflection. Never do I think of Coleridge, or turn a page of his, without feelings of affectionate veneration and gratitude as to a spiritual father. Wordsworth also I learnt to read at the same time. And I trust I am not violating the reserve of private history, which no one can prize more highly than I do, when I record, with grateful acknowledgement, that of all the instruction which it was my privilege to receive from Mr. John Campbell Shairp, now Principal of St. Andrews, this was not the least—that he led me to a knowledge of these two great men, directly to Wordsworth, indirectly and in conjunction with other influences to Coleridge.

Let me mention two of these influences;—first, the profound admiration felt for Coleridge by De Quincey, as shown by many passages in such of De Quincey's writings as were then known to me; and secondly, several expressions of Arnold's, especially this most remarkable one, which used to dwell in my memory. "His Table Talk marks him, in my judgment. . . . as a very great man indeed, whose equal

I know not where to find in England."[1] No *equal*, observe, in England, so far as Arnold knew. And a similar opinion held by De Quincey, a man in many respects so different from Arnold;—and both were names of highest authority with me.

The great Twin Brethren, as I will call them, Wordsworth and Coleridge, are the founders and inaugurators of the new era, the ninteenth century era, at least in England; Bentham and his school are but continuators of the eighteenth. Of the two, Coleridge still seems to me the greater. His mind is far more comprehensive, far more flexible, than Wordsworth's; he abounds in the prime gift of humour; Wordsworth is always either child or prophet, Coleridge is fellow man. And though doubtless the prophet's definite and concentrated force was the first to produce its visible effect on the age, still it does not follow that the slowlier working, and perhaps more subtil, force, diffused over a larger area, was in total effect the less rare and potent. Now Coleridge's writings appeal to those who require a logical diet. And somehow I found that he caught and arrested my attention, and made me ponder over passages to discover their meaning, when I should perhaps have credited no one else with having a meaning to be discovered.

For instance, where he writes : "It is a dull and obtuse mind, that must divide in order to distinguish; but it is a still worse, that distinguishes in order to divide. In the former, we may contemplate the source of superstition and idolatry; in the latter, of schism,

[1] Life and Correspondence of Thomas Arnold, D.D., p. 347. Sixth edition, 1846.

heresy, and a seditious and sectarian spirit."[1] Partly,
I believe, my attention was fixed here by what seemed
to me a hitch in the logic. A "still worse" mind
ought logically to mean, I thought, *intellectually* worse,
being contrasted with "dull and obtuse" minds. And
yet what followed seemed to imply that it meant
morally worse. What was the reason of this ?

Again, Aphorism VII.: "In order to learn, we
must attend: in order to profit by what we have
learnt, we must think—that is, reflect. He only
thinks who reflects." *He only thinks who reflects.*
Then what is the difference between them?

Many other passages I might quote, but these
are enough, and I have chosen two which contain
two distinct *sources*, so to speak, of the present system;—the principle of reflection, and the principle
of distinction of inseparables. Both come to me
directly from Coleridge, in the manner which I have
shown.

But after all, the great lesson which I learnt from
Coleridge was this, the intimate union between the
intellectual and the emotional elements in human
nature. Coleridge alone seemed to *know* what religion was, to know it, I mean, by *experience.* His
system, as a religious philosophy, provided for this
union. From this I concluded, not that his system
was true, but that no system could be true which did
not provide, as its outcome, such a scheme of the
universe as should not only enable, but assist, those
who held it to give the freest scope at once to their
intellectual and to their religious tendencies. Religion I saw was like an expansive force which would

[1] Aids to Reflection. Introductory Aphorisms, XXVI. Sixth
edition, Vol. I. p. 17.

shatter any man-made system of philosophy, unless that system were a true image of the universe itself. Nothing can be true which does not find a place, in the theory, for that passionate determination of the mind to God, which I do not say is described by, but which breathes from, the writings of men like Coleridge. And the reason is this, that the passionate religious tendency is not a sentiment fluttering round a fancy, but is a feeling rooted deep in the structure and mechanism of consciousness.

This structure and mechanism of consciousness it is, whatever be the facts with regard to its dependence on physical conditions, which has forced and is forcing us to shape ever new theories of the world, until we get the one which accords with facts, or in other words, in which all the facts accord with each other. And this view of matters is the gist and kernel of modern philosophy, and constitutes its *subjectivity*. No *external* test of truth can be even *sought for* any more.

But the emotions, and among them the religious emotions, are as deeply inwoven in the structure and mechanism of consciousness as any feature of sense or reason. They carry us deeply into the heart of things, the hidden springs of being, the inmost nature of the Existent. Now this is a doctrine which Coleridge, to use an expression of his own, did not possess, but was possessed by. He saw it as a seer. And indeed who does possess it yet, in any real sense? Who sees it otherwise than as a seer, or has knowledge of its length and depth and height?

But it is insight, not exactitude, that is the *life* of philosophy. Intensity in all things is of the feeling, not of the form. A living spirit breathes from Cole-

ridge's pages, which I at least can find in no others.
The world too has far more to learn from Coleridge
than it yet dreams of, not by way of system or theory,
but by way of vivifying impulse;

" To life, to *life* give back thine ear."

There is a scientific as well as a literary pedantry,
nor are we quite free from it in philosophy. Pedantry
is preferring letter to spirit wherever the two are at
variance, from the letter having been outgrown.
From this besetting sin of a learned class there is no
better preservative than to live in the genial atmo-
sphere of imaginative poetry. I speak only of poetry
which is the expression of imaginative emotion, not
of that of the fancy, still less of didactic poetry or
satire. The enjoyment of truly imaginative poetry
is incompatible with submission to any philosophy
that cramps the intellect. Its narrownesses are felt
to be narrow when you enter their cabin from the
open sunlight. Imposing subtilties and well-turned
phrases, potent in the study or the lecture-room,—
bring them out of doors, translate them if you can
into poetic language, force them to assume their true
shape. Do they still appear as truth, or have they
donned a semblance of jest or dreaming? If they
have, away with them to the limbo of fictions; in
philosophy, the land of comprehensive and yet inter-
dependent realities, they can at best be only half-
truths, and, if allowed to pass for truth, wholly mis-
chievous.

And this brings me to some remarks with which
I will conclude this Preface. Two questions there
are, of supreme practical importance at the present
time, on both of which philosophy throws light. The

first is this: Have we or have we not valid reasons
for conceiving of ourselves and the actual world in
which we live as surrounded by an *unseen*, but in its
nature phenomenal, world, of which ours is the seen
part, and with which it has real but unseen relations?
It seems to me that we have, and to show that we
have is one of the main purposes of the present
work.

The second question arises on this supposition.
Can we treat that unseen world, simply because it is
unseen, as if it were non-existent? Is it possible to
do so in the long-run? And if not, is it wise to endeavour to do so for a brief and uncertain time?
Surely it is not. The natural relations of man are
with infinity. Systems of philosophy, alike speculative and ethical, *must* recognise these relations. If
they do not, they are houses built on sand, which
the facts of human nature will lay in ruins;

    " Experience, like a sea, soaks all-effacing in."

In vain the Comtists, or any other school of
Positivists, construct philosophy without taking count
of the unseen world; in vain they build any system
of ethic on facts belonging to the seen world only;
when all is said and done in such ethical systems,
they find themselves either without a religion, or
compelled to invent one. To *invent* something which
is, as religion, to have binding power over its inven-
tors, a vessel of clay which the potter shall worship.
In truth, religion has no need to be invented; it is
based in the ineradicable sense of the relations of man
to the infinite, of the seen to the unseen world.

These questions, as I have said, are most important; it is so, because the answer to them must deter-

mine our whole scheme of education. Are we or are
we not to teach that there is an unseen world, and
that we have relations with it? The answer will
necessarily colour the whole scheme of the instruction given. Not only in schools, but from the first
dawn of a child's intelligence. And I maintain that
only *that* answer is the true one, only *that* system of
instruction will in the long-run be found practicable,
which is consistent with the largest view of nature,
that is, with the most comprehensive philosophy.
The motive force which lies in the conception of the
unseen world is an indispensable factor in moral
training. The question for teachers is, Will you
have this force for you or against you? It will assuredly shatter your systems of education, if you do
not include it in them. But to include it, you must
lay your foundations in the unseen world.

January 26, 1878.

BOOK I.

ANALYSIS OF ASPECTS.

CHAPTER I.

§ 1. DISTINCTIONS, not Definitions—such is, and must be, the primary basis of all philosophy. Before you can give a definition you must know in general what you are about to define, that it is something proper to be defined, and has a real local habitation in the world of thought. You cannot *begin* by defining; that would give only definitions "at large," *definitio vaga*. You must first have a guide to definition.

It is different in the case of some systems of philosophy. There the work of Distinction is supposed complete, and you begin with applying them to the phenomena; your country is already mapped, and you proceed to measure its divisions. Systems of philosophy which have not thoroughly done the preliminary work of distinction cannot be permanent. For instance, Spinoza begins with a definition of Causa Sui; "By Cause of Itself I understand that, the essence of which involves its existence; or again, the nature of which cannot be conceived except as

---

[1] This Chapter appeared originally, in three parts, in the three first Nos. of MIND, Jan. to July 1876. Some slight alterations and additions have been made.

existing." Very good; but *is there* such a thing?
Is such a thing possible to thought? There is at least
one term here which calls for analysis. "Essence"
may be considered to be sufficiently explained by
being distinguished into the *nature* of anything as it
is *conceived*. But Existence, what is that? Till we
know that, we are ignorant whether any essence can
possibly involve existence; whether putting "existence" into the definition of anything makes that thing
to exist. There is a good deal of distinction-work to
be done with reference to "existence" before a causal
connection between a thing and itself, *causa sui*, can
be founded on a conceptual connection between the
essence and the existence of that thing. Till then,
the famous definition of Causa Sui is all in the air, a
definition "at large."

System then or no system, the first thing to be
done and done thoroughly in Philosophy is to distinguish,—to distinguish in order to know what to
define, and what sort of notions to employ in defining
it. Let us see you handle the subject, turn it about,
exhibit its facets; let us know *what sort of a thing*
you imagine it to be, where are its boundaries, who
are its neighbours, what its products,

> " Arvo pascat herum, an baccis opulentet olivæ,
> Pomisne an pratis an amicta vitibus ulmo :
> Scribetur tibi forma loquaciter et situs agri."

The first distinction to be established, and one
which is a pre-requisite of all the rest, is between
philosophy and science ; the ground must first be
won before we can proceed to distinguish the several
provinces which it contains; there can be no distinctions within philosophy, unless there is a philo-

sophy which is itself distinct from all other branches and kinds of knowledge.

This distinction cannot be a total separation; an unscientific philosophy would be no philosophy at all. But the distinction may be drawn in many ways, of which only one can be the true one. Four ways of drawing it may be enumerated as follows:

First, it is possible so to draw the line between them that nothing remains for philosophy but the preliminary guesses at truth which men have made before striking into the true methods of discovery, which true methods with their results are science, and supersede the old mistakes, which are philosophy. If this were the true account of the matter, philosophy would have no *locus standi* in the intellectual world, only the ignorant would be its votaries, and philosophers would be no better than obscurantists, basing themselves more or less consciously on the maxim, *populus vult decipi et decipiatur*. This way of looking at the matter, being very prevalent in England, may perhaps be called English Positivism.

Secondly, the line may be drawn between them by saying, that as science advances, and divides into many branches, room is made for a coördination and systematisation of all, which is a work demanding separate treatment and separate labourers; and that this work is philosophy. This view is that of Comtian Positivism.

Thirdly, it may be maintained that philosophy is the discovery of Absolute Existence, and that the sciences only then become scientific when they are deduced from the laws of this absolute existence, from which they thus receive their whole scientific character. This is the Hegelian view.

Fourthly. a position may be taken up which ascribes to philosophy as its special work, besides the coördination and systematisation of the second head, a negative task.—the task of disproving and keeping out of science all ontological entities, whether these appear merely as spontaneous products of the uncorrected imagination or have been reduced into systems, such as for instance the Hegelian. This view is that taken by Mr. Lewes in the important work[1] which is now in progress. Mr. Herbert Spencer's view also, which runs up all knowable phenomena to a foundation in the Unknowable, is very similar.

There is yet a fifth view possible. the one which I shall endeavour to establish in the present Chapter. Briefly stated, it is this: Philosophy is more than the coördination and systematisation of the second head, and more than the negative function of the fourth head; it has a positive content and a positive method of its own, and yet a content and a method which are in no sense ontological or transcendent. And this method and content are the permanent and indestructible *raison d'être* of philosophy, assuring to it an existence as a distinct kind of science.

Let me be allowed to dwell a little on what is involved in this view, which I have stated at present in very general terms. If philosophy has a distinct method and a distinct and positive content, it follows that there has been for some definitely assignable period a growing system of philosophical doctrine, of philosophical truths retained distinguishable from philosophical errors discarded. a system due not to one or two philosophers only but to many, the

---

[1] Problems of Life and Mind. See particularly Vol. I. pp. 62. 75. 86 ; and Vol. II. p. 221.

growth not of a single epoch but of centuries. There must be a history of philosophy different from the history of successive systems of philosophy, and from the law of their succession. The systems of philosophy are not philosophy, its history is not the history of their succession. It follows likewise that there cannot be a history of philosophy until the object of that history, philosophy itself, the growing system, has been detached and delineated.

But what meets us most prominently, when we first turn our attention to philosophical subjects, is the apparent absence of a philosophy, the obvious presence of a multitude of conflicting systems. What is the explanation of these two facts? The readiest explanation is offered by the first of the views enumerated above; the systems are present because undisciplined minds have abounded, the philosophy is absent because it is non-existent. But on the view which I am about to maintain, this easy explanation of the facts cannot be the true one. The true explanation is, that philosophy is apparently absent because it is yet in its infancy, and the systems are obvious because they are necessary means of giving it birth. The systems would on this view have served a purpose consistent with their own untenability, and philosophy would have been receiving form independently of their decay. It is true that on this supposition philosophy must be as yet in a very early stage of its development, and so no doubt it is. Its systematisation as an organic whole is most imperfect; organisation is its primary need. But everything seems to me to show the possibility of such an organisation, the possibility of marking out and giving coherence to a body of philosophical doctrines which shall form for philosophers of all schools a common possession and a common basis, as they will assuredly have been won by a common effort.

Nevertheless system-making in philosophy cannot be laid aside; there is one indispensable function which it alone can perform. It is the mode by which verification is effected; it is to philosophy what verification by observation and experiment is to the physical sciences. And by the nature of the case it is the only verification of which the phenomena of philosophy are capable; for these are not like those of the physical sciences, things which fall under the cognizance of the outer senses, but pure representations—pure, that is, from presentation; with these science ends, and with these philosophy begins. Its theorems consist not simply in thoughts about things, but in thoughts about thoughts of things. These pure representations however which are the phenomena, the facts, of philosophy, must always be verifiable by the facts of nature, that is to say, in technical terms, by the presentations which they represent. In many cases these verifications are so simple that any one can perform them without a special scientific training, as for instance in the pure representation 'all visible surfaces are coloured.' Others are more difficult, and here we must have recourse to science to prove the truth of the representation before we can admit it as a fact in philosophy.

Thus the law of gravitation is, in science, a thought about things, being, in nature, a general fact in the things themselves. Here the verification consists in examining the things. But the law of gravitation as it is in Science, in its character of a thought about things, becomes in Philosophy the object-matter of a further examination, a philosophical one ; it becomes one of the phenomena of philosophy, and the basis of thoughts which have thoughts of things for their object. Here the verification of any theorem of philosophy relating to the law of gravitation must consist not in comparing the law of gravitation with physical phenomena, which is a verification belonging to science, but in comparing the theorem of philosophy with the law of gravitation as it is in science. The ultimate as well as the particular laws of science are among the phenomena of philosophy; it is only to be regretted that they are still so few. While, then, the laws of science are verified by the facts of nature, those of philosophy are verified by the laws of science; in other words, theories of philosophy must be made to harmonise with the laws of science so far as these are at any time known; and it is from this requirement that all legitimate system-making in philosophy springs.

In these remarks we may also read the explanation of the predominantly literary character of philosophy in contrast with science, of its workshop being the library not the laboratory, its pabulum the writings of previous or contemporary philosophers. For philosophy is primarily and mainly, I mean in its whole analytic branch, concerned with *clearing the ideas*, not with discovering new facts, but with analysing old ones; its problem being not how the world came into being, but how, having come, it is intelligible.

I now proceed to establish the true distinction, as I conceive it, between philosophy and science. In the first place it is abundantly clear that they have points of agreement. Going back to the meaning of those who first called themselves philosophers, lovers of knowledge instead of possessors of it, it is clear that the position which they thus took up was not one of disregard to knowledge already attained, to knowledge in and for itself, but the adoption of a new point of view by the observer towards that knowledge; it involved a generalisation of the notion of knowledge, and brought out the fact, that while they were possessors of some portions of knowledge, they were only aspirants to possess other portions, which other portions were to them as yet unknown and only to be called knowledge *in potentia*, *in futuro*. At the same time this future, and not yet actual, knowledge was necessarily assumed as being of the same kind, in point of being truly knowledge, as those portions which were already reduced into possession. Philosophy, then, was conceived as a further search, a pioneering expedition, into realms as yet unknown, in order to bring them under laws of the same kind as those which constituted the knowledge already discovered.

So far there is, it may be said, no very wide distinction between philosophy and science; for science too must always have recognised the search for further knowledge as essential to itself; a science which professed to contain only what was already known, and not also means and methods for future discoveries, would be a mere *scientia docens*, not *utens*; and philosophy would be merely a grandiloquent name for one part of science, for that part of it which faced forwards into the as yet unknown and undiscovered. In short, if this distinction were all, the first of the views enumerated above would be fully justified.

But now comes another distinction. As science advances, her discoveries are made piecemeal, one by one; as they are made they are compared and classified; and thus along with the general advance of science there goes on a distinction of the whole into special sciences; and as the number of new discoveries increases in each branch of science, the growing mass and complexity of each branch becomes sufficiently great to occupy and more than occupy the whole energies of individual men, leaving them no disposable opportunity for making discoveries in other branches than their own. But in every special branch of science, as it is thus called into being by the growth and development of knowledge, the same distinction prevails, I mean the just noted distinction between present and future knowledge, between hypotheses that have and hypotheses that have not yet been verified. Here it is that the distinct scope of philosophy takes, as it were, a second step towards its manifestation. And the general forward outlook in the special sciences taken together, as distinguished from the already acquired knowledge, taken together, in all of them, is that which marks philosophy in this its second, but still most rudimentary, stage of distinction from science. Philosophy appears in this second stage of its life, so to speak, as taking the results acquired by each of the special sciences, and endeavouring to frame hypotheses which should unite them into a single system, and make them serve as a guide suggestive of new hypotheses.

The rudiments of the notion of philosophy, as distinguished from science, are thus given by the two combined characteristics of generality and hypothesis. But the rudiments only. And these same characteristics contain in themselves the germ of a third, which is necessarily developed from them. If we stopped short at these two, seeing nothing else in philosophy to differentiate it from science, we should find ourselves holding the second view, that of Comtian Positivism. For it may be argued that, even supposing the greatest completeness in the number and organisation of the special sciences to have been reached, and by consequence the greatest generality in the hypotheses which will connect their results into a system of the whole; in which case the greatest possible difference would exist between the functions of science and those of philosophy, as they have been up to this point delineated ; even then, it may be said, the functions of philosophy, so far as they have any scientific value, are not different *in kind* from those of science. Philosophy, the framer of general hypotheses, is merely a special science to which a particular task is assigned, for convenience sake, that of coördinating the several sciences into a single system of sciences, and the results of all into a single science of nature. The main problems of philosophy would be two, or rather one with a double aspect, the Classification of the Sciences, and the Codification of the Laws of Nature; in fact just what Comte aimed at in his first great work, the *Cours de Philosophie Positive*. But neither of these problems is different in any essential characteristic from those of science proper, that is, from science in any of its special branches. The distinction of philosophy from science would be, then, in this case a detail, most important it is true, and even necessary, but one resting on no fundamental difference in their functions.

All this I take to be indisputable; and if no other distinction than the two already mentioned can be shown to exist between philosophy and science, then it must be admitted that philosophy has no special *raison d'être*, no claim to a separate and independent but merely to a nominal existence, such as the term Positive Philosophy is intended to accord to it. I proceed, then, to show that there is a third characteristic, by which, in combination with the two former ones, philosophy is distinguished as different in kind from science.

All the special sciences, in their demonstrations, run up to certain ultimate notions as their basis of demonstration, and there they stop. Beyond these they do not care to pursue their analysis, content with the acknowledgement, which no one refuses, that those ultimate notions which they take as their basis correspond to realities of experience, and represent those realities with essential accuracy. Some among the special sciences base themselves upon notions which they take from other special sciences more abstract and more general than themselves; physiology, for instance, partly upon chemical notions, partly upon mechanical, partly upon physical; chemistry bases itself partly upon mechanical, partly upon physical; these two last run up again into what is called Rational Mechanic; and here for the first time we meet with ultimate notions which are not derived from any other more abstract special science, but are drawn directly from the fountain head, experience.

These ultimate notions are Mass, and Energy Potential and Kinetic. That is the shape into which Rational Mechanic has thrown the two older and vaguer notions of Matter and Force, for the sake of first defining them and then exactly calculating or measuring them. Mass is measureable matter, 'quantity of matter' being its definition. Energy, potential and kinetic, is phenomenal and measureable force, as distinguished on the one side from force as the cause of motion, on the other from particular forces, that is, groups or modes of motion of a particular kind, as for instance gravitation, or electricity. For both force and energy involve the notion of motion, the motion of masses or portions of matter in action and reaction with other portions. And both in mass and energy taken together, and in matter and force taken together, motion is involved. Motion itself again is abstracted and treated apart from the different kinds of matter which move, in a separate branch of science known as Kinematic; and this branch forms the connecting link between rational mechanic and the sciences of pure mathematic. What I have, then, specially to observe is, that in rational mechanic we meet with elements or notions which are not derived from pure mathematic, and which have no other source than direct experience; and of these notions, which in their most abstract and general shape are called Matter and Force (measureable and calculable under the terms Mass and Energy), science can give no other account than that they are facts, and ultimate facts, of experience. Experience is their source, and experience also furnishes the verification of the reasonings concerning them.

Rational Mechanic, in respect of its other elements, holds of Geometry and the sciences of mathematical calculation, Arithmetic, Algebra, and the

Calculus, through the medium of Kinematic. And
these sciences include between them, and are based upon, the notions of abstract Motion (which involves those still more abstract of Space and Time), Number, Quantity, Continuity, Discontinuity, Infinity, and Figure. Pure Mathematic includes all the methods of calculation and measurement so far as they are irrespective of what the things are which are calculated or measured. And as such these sciences base themselves upon certain ultimate notions which serve as principles of the process of calculating and measuring.

The question accordingly arises with respect to these sciences of pure mathematic,—Are they competent to explain thoroughly the nature of those notions which they assume as their ultimate bases of demonstration? Does for instance the calculator explain what an Unit is? Certainly not. All he tells us is— We can count anything *once*. This *once* is the unit of numeration, and it is obviously independent of, and indifferent to, any particular kind of object counted (or measured) by it. In fact, he *defines* an unit, and defines it sufficiently for his purpose; it is defined in such a way as to serve for a basis of further reasoning, but not in such a way as to show on what it is itself based. He *defines* but does not *analyse* it.

Again, does the geometer explain how and whence he comes by his object-matter, how he comes to regard pure spatial extension as figured? No. He *begins* with figured space. Either he begins with the notion of Volume, and proceeds to analyse it by the ways in which it is bounded, or else he begins with the notion of Boundary, points, lines, surfaces, and proceeds to the construction of Volume. The Configuration of

Space is his object-matter; and he analyses this, notionally as well as actually, to its remotest part; but he assumes Figured Space, in the general, *as a datum*; he does not tell us how it comes to be possible, but contents himself with saying that we all know it to be so, and that this his basis is sufficiently clear in meaning and secure in reality.

As I am not primarily occupied with the interconnection of the sciences, it will not be expected that I should have stated the exact moment at which these ultimate notions are introduced into the sciences, or have made a distribution of them beyond the possibility of objection. It is enough that the positive physical sciences between them, from physiology to mathematic, do introduce these to them ultimate notions, namely, Mass and Energy (which may be taken as involving the higher notions of Matter, Force, Cause), Motion, Unity, Length of Time, and Configuration of Space. And I think I have made it sufficiently evident, that these ultimate notions, ultimate to the physical and mathematical sciences, are not ultimate in all respects. They are ultimate in respect that we can securely reason downwards from them, that is to say, construct valid definitions of them, and base valid demonstrations on them, in the physical and mathematical sciences; but not ultimate in respect that we can analyse them still farther, reasoning upwards from them, and ascending to still higher generalities and greater abstractions. Their validity as the basis of science is sought and found in what lies below them, in the concrete objects to which they are to be applied. It is conceivable they should also have another validity as deductions, or cases, of higher abstractions, to which they in their turn would serve as a basis of validity and as concrete object-matter.

The question whether any such higher abstractions are discoverable is thus posed by the sciences themselves; and the conditions of its solution are also laid down in the posing. We are required to find an answer to the questions, *What are* Mass, Energy, Matter, Force, Cause, Motion, Unity, Length of Time, and Configuration of Space? And the conditions of solution are, that the answers shall be in terms which do not repeat again the things about which the question is put (the common logical requirement in all solutions), but shall consist of higher generalities or abstractions, which yet shall be really known to us (not fictitious), and shall thus present us with new knowledge about the things in question. In other words, the notions in question are to be analysed or resolved into elements more abstract than themselves, which elements, in composition, shall give us again the original notions.

Now in thus approaching the question whether any such higher abstractions are discoverable, every way but one is barred to us. We start from notions representing concrete objects of experience, and representing those objects already in the most general and abstract shape. We cannot therefore look for the answer in any objects of concrete experience, or in notions representing them; because this would be to go to notions less, instead of more, abstract and general. We must pass beyond all concrete objects of experience, and beyond the most general notions which we can frame of such objects; and we have to answer the question *What?* τί ἐστι; concerning these most abstract notions. Where, then, is there a limit to our thought within which we may have been confined consciously or unconsciously,—a limit which is now to be removed and give freer scope to thought; where has there been a restriction which it is possible to take away? If there has been no such limit, no such restriction, then we cannot take a step beyond where we are already; we are already at the end of our tether, and *every* road is barred to us. The ultimate notions of science are then for us the ultimate notions in every respect, and the question whether we can refer them to higher generalities is answered in the negative.

But it becomes clear on a little further reflection that there has been such a limit and restriction, a limit by removing which we can take a step in advance and reach a still higher generalisation, yet without passing into the region of fictitious entities. For we have hitherto been regarding the objects of our enquiry *as objects*, that is to say as endowed, some way or other, with an existence independent of ourselves the spectators of them; or, if we have made a reservation to the effect that these objects are, after all, only phenomena relative to the percipients, still we have not as yet made any use or application of the reservation. But now the moment is come at which the fruits of the reservation may be reaped. We find that we can analyse the ultimate notions of science still farther, by looking upon them as phenomena relative to the percipients, and asking ourselves what features they possess in this their *subjective* character, in their character of states of consciousness as contradistinguished from their character of objects, or portions of an objective world. We are thus simply taking the obverse aspect of the very same ultimate notions which we were dealing with before;

and the result is a new, and subjective, analysis of
those notions which in their objective aspect (in
which they were the bases of the sciences) appeared
to be unanalysable and ultimate.

It is found, on thus regarding them, that certain
modes of Sensation in combination with pure spatial
extension and pure time-duration, are the constituent elements of each of these ultimate notions taken
subjectively. And by pure spatial extension, and
pure time-duration, I mean the space *element* and the
time *element*, in and with which any sensation is felt.
Every sensation without exception has a time element; every sensation of sight and of touch has a
space element as well. And by calling this element
pure, I mean that it is different from the sensation,
and as different from it is unaffected by division,
continuous, having no divisions of its own, but receiving them from sensation. The divisions of pure
time and of pure space are given only by changes in
sensation, and without these divisions of pure time
and pure space we should have no consciousness
whatever of time in lengths of duration, or of space
in its configurations or relative positions of points,
lines, or surfaces.

We have also here the source of the notions of
continuity and discontinuity; of quantity, which is
the sole object of measurement; and of infinity, the
notion of which is nothing but continuity without
break, or abstracted from discontinuity. Number
rests upon the notion of unity, or of an unit.

To count a thing *once*, which is the notion of an
unit, depends on that thing being distinguished by
change of sensations from what precedes and follows

it in consciousness, no matter whether that change is arbitrarily introduced by ourselves, as in the case of units of measurement, or not.

Motion requires change not of sensations simply, but of their position in space, taking place in succession of times.

Cause involves the notion of the inseparability of things previously regarded as separable. But to treat things as separables is to treat them *as if* one was before the other in time, whether their order of sequence may, or may not, be equally well reversed, and the things found to be simultaneous in time. Cause therefore requires the notion of sequence of sensation.

For the notion of force (if it is held necessary to introduce it into science in the character of a cause of motion), a peculiar class of sensations is required, that of muscular tension or effort, whether derived from efforts of our own which we feel ourselves, or from these carried over in imagination and attributed to objects which are or may be in opposition to ourselves.

Energy, if not explained by reference to force, is in that case simply a derivative of motion. It consists of changes in the position and motion of masses and parts of masses.

Mass, as remarked above, is nothing but matter scientifically treated.

And lastly, that solid resisting thing which we call Matter requires for its comprehension (speaking only of normal cases) sensations of sight in combination with those of touch and muscular tension. At any rate sensation (whether of sight, or touch, or both combined), but always in spatial extension, is the necessary and sufficient analysis of our notion of Matter.

It must suffice, in a Chapter like the present, just summarily to indicate the nature of the questions and answers which arise on passing onwards from the ultimate notions of science to their analysis as states of consciousness. As above I could do no more than enumerate the ultimate notions of science, without attempting to assign them with perfect accuracy to their respective places in science, so here I must content myself with indicating, and cannot pretend to demonstrate, the general nature of the analysis which these notions receive in philosophy. That analysis is a final one, in the sense that there is no further conceivable limit the removal of which would throw open another field, as the removal of the objective limit unbarred the entry into the field of subjectivity. The analysis is also an analysis of the *nature* of the things analysed, not on account of how they arise or what are their antecedents. Ultimate subjective analysis of the notions which to science are themselves ultimate,—such is the answer which I have to give to the question, What are the features which distinguish philosophy from science?

Up to this point, it will be observed, we have been occupied with the relation of philosophy to one class of sciences only, the physical and mathematical. When we come to the other classes into which the sciences are usually, and exhaustively, divided, a similar conclusion will be forced upon us. A *similar* conclusion, because in these classes of sciences, the Moral and the Logical, the ultimate notions which are their distinguishing and characteristic marks are already subjective; for which reason it is that these sciences are most usually treated as forming a part of philosophy as distinguished from science.

Interwoven as all the moral sciences are at every step with those of the physical and mathematical series, yet their subjective character is everywhere predominant, and their objective subsidiary. They are *practical* in their character, that is to say, the comparative importance of motives to conscious beings, the comparative value of states of consciousness, is the chief matter of discussion and enquiry. Whatever notions we take as ultimate in any of them, whether (for instance) that of Justice and Injustice in Jurisprudence, of Wealth in Political Economy, of Beauty and Deformity in Æsthetic, of the Good of a Community in Politic or Sociology, of Good and Evil in Ethic,—these ultimate notions, ultimate in respect of the particular branches of science which are based upon them, are yet capable of a further analysis into elements, an analysis not indeed differing from what has preceded it in point of subjectivity, since both alike are subjective, but still an analysis more searching than would be strictly necessary for a definition which should afford a basis for a branch of science. I mean that, with less searching analysis and consequently less accurate definitions, the sciences based on them would be less perfect, but not therefore impossible.

In Logic again we have, as its ultimate basis, the three Postulates known as the laws or principles of Identity, Contradiction, and Excluded Middle. Upon these the whole doctrine of Logic rests, and for its validity no more is requisite than the statement of them. They carry their evidence in themselves. They are in a precisely similar position to that of the ultimate notions of mathematical science. They have, too, as being even more abstract than most, if not all, of the latter notions,—they have immediately attaching to them the double attribute of subjectivity and objectivity. They are at once laws of things and laws of thought. At least if they should be finally held *not* to be immediately laws of things, the discussions which have been raised upon the point suffice to show the *appearance* of such a double character in them. But even in their case a further subjective analysis is possible, an analysis by no means requisite to assure us of their validity, but certainly requisite to ascertain their nature. This analysis is of the same general character as in the case of the ultimate mathematical notions. It is into some particular *volition* and *time;* that is to say, we must attend to some feeling, distinct from others, before we can say, This feeling is this feeling (A is A); This feeling is-not what is not this feeling (No A is Not-A); and Everything is either this feeling or what is not this feeling (Everything is either A or Not-A).

The several sciences then, in every case, yield us notions, their ultimate bases, which are susceptible of a further subjective analysis, whether these notions are themselves objective as in the physical and mathematical sciences, subjective as in the practical, or both at once as in logic. But besides these ultimate notions of the several sciences, there is yet one notion to be mentioned, a notion not peculiar to any one science, but common to all, and involved in the particular ultimate notions of each. This notion is that of Existence. Different as the three groups of sciences, physical, logical, and moral, are in point of

subjectivity and objectivity, yet the notion of Existence is involved alike in all. Not Matter only but
States of Consciousness also have existence; they are
what they are and while they are. What, then, is
the notion of Existence, and does it belong to science
or to philosophy to answer this question? It clearly
belongs to philosophy; first, because the notion of
existence is more general and abstract than any of
the ultimate notions of the physical or mathematical
sciences; and secondly, because subjective existence,
a notion which emerges first in philosophy, is an included part of the general notion which embraces
existence both subjective and objective. We may
put these two reasons in somewhat different phrase.
The subjective aspects of material objects exist, as
well as the objects themselves; and states of consciousness, such as are the emotions, and feelings of
pleasure and pain, which have no material objects,
yet exist for the Subjects of them.

Subjective states and objective things, then, are
both alike *existents*. But they stand in a somewhat
different relation to consciousness. The objective
things are the nearer of the two to the consciousness
both of the individual and of the race, counting from
the moment when he or it begins to philosophise;
the subjective states are the nearer to the consciousness of both, counting from the epoch when sentience
arises. We begin to philosophise having "objects"
already formed in the mind; but there has been a
·process by which these objects have been formed,
prior to philosophical consciousness, but not prior to
consciousness generally. It is a case for the application of the maxim—What is last in analysis is first
in genesis; and, What is last in genesis is first in

analysis. Thus it has long been observed and often repeated, that the distinction between the two kinds of existents, subjective states and objective things, is not perceived at the earliest stage of an individual's experience.

> " The baby new to earth and sky,
>   What time his tender palm is prest
>   Against the circle of the breast,
> Has never thought that ' this is I :'
>
> " But as he grows he gathers much,
>   And learns the use of ' I,' and ' me,'
>   And finds ' I am not what I see,
> And other than the things I touch.' "[1]

When, however, this distinction is perceived, then both kinds of existents become objects to the percipient; and the perception of both, in their contra-distinction, is itself distinguished by the name of reflective perception as opposed to direct, and by that of self-consciousness as opposed to consciousness simply. It is this "moment" of reflective perception or self-consciousness which is the central and cardinal feature in philosophy, and that which, by enabling us to distinguish the subjective from the objective aspect of things, distinguishes philosophy from science by an inner and indelible characteristic.

The answer therefore to the question, What is Existence? can only be given, if at all, by philosophy. But what that answer will be I am not now to discuss. In general terms it may be said that, for philosophy, existence means presence in consciousness, *esse* means *percipi* ; and this quite generally, so as to include all the modals into which the general proposition may be thrown ; as for instance, possible

---

[1] Tennyson, In Memoriam, XLIV.

existence designates what is possibly present in consciousness; actual existence what is actually present
in consciousness; imaginary existence what is imagined as present in consciousness; necessary existence what is necessarily present in consciousness; and
so on. For all the modes of existence there are corresponding modes of presence in consciousness, and
without a corresponding mode of presence in consciousness we should have no knowledge whatever of
any mode of existence,—neither what it was nor that
it was. In short, consciousness itself is the subjective
aspect of existence, and each in its bare generality is
the ultimate and common feature of which all the
modes of consciousness on the one side, and all the
modes of existence on the other, are differentiations.
In this most abstract and general character, their
character as *summa genera* of Modals, they are unanalysable into elements, consequently undefinable,
and only so far capable of explanation as the two
unanalysables, the two opposite aspects, existence
and consciousness, throw mutual light on each other.
We know existence as consciousness, and to know
that we do so is self-consciousness.

§ 2. I have now to advert to a peculiarity which
will open up an entirely new branch of the subject.
The distinction which has been established between
the subjective and objective aspects, and which is the
basis of that between philosophy and science, is one
which rests on the support of no previous theory as
to the substratum or agent of consciousness, any more
than it rests upon any theory as to a corresponding

substratum of Matter, or generally of the objective aspect. No soul, or mind, or ego, or nervous organism, is assumed as the thing which *has* the states of consciousness. No material existent is assumed as the thing which has the properties of resistance or impenetrability, or which is the seat of the forces by which matter is actuated.

On the contrary, and this is the point now specially to be noted, the distinction between the subjective and objective aspects precedes, and is required for the formation of any such theory, of whatever character it may be, relative either to Mind or to Matter. This will be clear if we reflect that, before we can devise an hypothesis to account for the existence of either aspect apart from the other, we must have distinguished, however roughly, the two aspects themselves.

Now it will, no doubt, have occurred to readers who have followed me up to this point, that there has been an important omission in my enumeration of the sciences which run up into philosophy. I have omitted all mention of the science of Psychology. This omission I am about to rectify. Psychology has all states of consciousness for its object-matter; and so far it has precisely the same object-matter as that here attributed to philosophy. Now psychology is a science, and that science which is the peculiar glory of Englishmen, having been, if not created, yet chiefly cultivated by them. It would seem then that, by simply adding the science of psychology to the list of the other sciences, we cover the same ground and perform the same service as we should do by superposing philosophy on the sciences, as something generically different from them. One or the other appears superfluous, and in such a case the simplest expedient must be the best, and philosophy must give place to a less pretentious rival.

It is here that the remark just made finds its application. The main purpose of Psychology is to investigate the laws by which different states of consciousness either co-exist or follow one another; it leaves behind it the mere analysis of particular co-existences and of particular sequences of conscious states, and, by comparing several instances of them, endeavours to discover the general laws which connect particular states into sequences or into co-existences. It seeks the conditions of their appearing in this or in that connection. Leaving their mere analysis, which assigns their elements of analysis, their nature, or their conditions *essendi*, it seeks their conditions *existendi*, that is, their genesis and history. It assumes them, therefore, to be not only distinguishable, as in analysis, but also separable, capable of existing as parts in different connections or wholes. It starts from states of consciousness as units, not indeed necessarily capable of existing alone, but still units capable of entering into various combinations.

But this search for the laws, or relations of dependency one on another, between states of consciousness is at once guided by facts to the objective aspect of the states of consciousness, excluding their subjective aspect. It is "things" outside the body which appear to cause "subjective states" within the body. The search for laws of dependency forces us not only to separate the states of consciousness from one another, but also to separate states of consciousness as subjective from their objective aspect, that is, from the same states of consciousness as objective, in other words to separate Subjects generally from Objects generally.

For relations of dependency have in all other sciences been found to exist only where the thing from which the dependence moved, that is, the condition or cause, was of a solid and material nature, a substance, capable of existing ἐνεργείᾳ. Psychology therefore, in seeking the conditions *existendi* of subjective states, seeks them in the laws or in the nature of substances, only reserving the question whether there is, beside the organism and the objects external to it, a substance residing in the organism but of an immaterial nature, that is, a Soul or Mind. Psychology passes in this way beyond the field of mere subjective-objective analysis, and envisages the particular relations of dependence which particular portions of the subjective aspect have to particular portions of the objective. And it is therefore not permitted, like philosophy, to abstract from the substrate or agent which has the states of consciousness; for it is only in and by such a substrate or agent that the causal nexus in its sequences and the dependence in its co-existences can be accounted for.

But if this is the distinction between philosophy and psychology, the question immediately arises— May not philosophy, then, be regarded as a part, the analytical part, of a larger whole, psychology? There are two main reasons against so regarding it. The first is drawn from another application of the remark above made. To do so would involve an inversion of the logical and historical relations between the two. *Historically* there was the germ of a philosophy, a distinction between the objective and subjective aspects, before there was the germ of a psychology, an enquiry into the conditions of existence of the phenomena of the latter. And *logically*, the distinction of the aspects is the prior condition of the enquiry ; for distinction must precede separation, and, as we have seen, it is psychology that first separates the two aspects; in doing which it gives back, as an object of direct consciousness, things which were in philosophy the object of reflective consciousness.

Here we come to the second reason. The analysis of states of consciousness as given in philosophy takes those states in connection with their objective aspects; these objective aspects it is which give us the states to be analysed ; but in psychology it is in reference to their conditions in the organism, or other substratum, that they come under analytic dissection. The former is a general, the latter a special, method. There is a common object-matter for analysis, namely, states of consciousness, in both ; but in philosophy we look for features which reproduce the world at large, in psychology for features which we can connect, as dependents, with qualities or properties of the conscious organism, or other substrate of consciousness; disconnecting them from their objective aspects in the world of existences, and thus assuming the *separability* of the subjective and objective aspects. And necessarily so, for we are here occupied with this question among others, how far the subjective states of consciousness are a correct image and reproduction of the objective world. But when we take these same states of consciousness in philosophy, we disconnect them from their conditions in the conscious organism, and connect them with their objective aspects in the world of existences; thus assuming the *inseparability* of the subjective and objective aspects. And we are enabled to do this without danger of erecting subjective fictions into truths, because in philosophy we do not begin with the subjective aspects but with the objective ; we take the ultimate truths of the sciences and enquire what are their subjective aspects, and do not take any supposed ultimate subjective aspects, and ask what their objective aspects, what their corresponding existences, must be. The method and assumption of philosophy are, in this sense, diametrically opposite to those of psychology. It is a different but perfectly legitimate way of looking at the same phenomena, though in so looking at them they assume a different complexion and give rise to a different set of distinctions and definitions.

I argue therefore that it is not permissible to classify psychology and philosophy, so opposite in point of method, so different in point of object-matter, as parts of a single science ; and still less permissible to call philosophy the analytical part of a larger whole, psychology, seeing that philosophy is not only prior in logic and larger in scope, but also has a method corresponding in generality to its larger object-matter.

For let us consider for a moment what it is that constitutes a separate science, and demarcates one science from another. It is not merely an arbitrary difference in point of object-matter; nor yet is it an arbitrary difference in method; but it is the mutual determination of method, in the first instance, by object, and then of object, in the second instance, by method. There is no science of the individual, nor yet of any individual class of things. It is always a general feature or features which is the object of a science. The same individual things are the object of Mechanic by reason of displaying the general feature of potential and kinetic energy, and the object of Chemistry by reason of displaying the general feature of molecular affinity in composition and decomposition. Wherever any general feature is such as to be accessible in a particular way better than in others, that way of access is the method of the science, and that general feature, wherever found, is its object-matter.

Physiology investigates the general feature, Life, wherever found; that is, in living organisms of all varieties. Psychology investigates the general feature, Consciousness, in living organisms, that further feature in them not investigated by physiology. The range of psychology is an enlargement of that of physiology, for only objects in local contact with the organism directly influence its vitality, whereas things not in local contact, but imagined only, may be said to influence its consciousness, and indirectly its vitality, —such merely imagined things being the index and evidence of nerve-processes which at once subserve consciousness and are endowed with vitality.

Psychology, then, differs from physiology in this, that it brings in subjective states as part of the general object, Vitality of organisms, and thus gives a new complexion to the phenomena of vitality; it has the old object-matter with additions, and therefore in a new shape. For its method it depends partly on Reflection; as we have seen above, that the subjective aspect must first be *distinguished*, before it can be *separated*, from the objective. But psychology is not the first science to make this use of Reflection, to adopt and employ the distinction of subjective and objective aspects. All the other sciences require it in the same way; the difference is, that they bring into their object-matter portions of the objective aspect only, *i.e.* Things, the external world;

whereas psychology brings into its object-matter subjective states as such.

But what most decisively distinguishes philosophy from psychology, as well as from all the other sciences, is its elevation of Reflection into a method. And this elevation introduces a new feature into the general object-matter, namely, the feature of inseparability of the two aspects. They never were, in fact, separated; but this fact had not been adverted to. To advert to it, to become aware of it as a general truth, is to elevate the act or process of Reflection into a method. In employing it we continually ask what we mean by such and such terms, what is the analysis of such and such percepts. We have thus a method which is all-embracing in its scope, for there is no word, no thought, of which this question may not and must not be asked.

While therefore philosophy is a further differentiation of the general object-matter of psychology and the other sciences, it is also a new method, and the method corresponds to the differentiation. Method and object-matter together make it a separate science, demarcated from psychology very much as psychology is from physiology. Some perhaps there are who would class psychology as a part of physiology, or both as parts of biology. But however we may class them nominally or for occasional convenience, the difference of method, mutually determining and determined by the difference of object-matter, is that which it is practically as well as theoretically important to observe and retain; for it is this which constitutes the permanent articulation of the scientific system, and this by which it corresponds to the distinctions of nature. On this ground therefore I con-

Book I.
Ch. I.
―――
§ 2.
Special relation
to Psychology.
tend, that philosophy is demarcated from psychology
by a difference as permanent and complete as that
which demarcates psychology from physiology, or
any one of the special sciences from the rest.

Let us now cast a glance at the practical bearings
of the subject. Philosophy, it has been maintained,
is not a part of the larger whole, psychology, in point
of theory at least. Is there any reason for treating
it in practice as if it were so? If there is, it must be
based on the fact that a better and more searching
analysis is afforded by treating philosophy as a part
of psychology than by taking it separately and then
making it an independent ally. I maintain that there
is no better but a worse result on the whole to be
anticipated from pursuing the two as if they were
one, and that one psychology, than from pursuing
the two independently and using each to correct and
control the other.

The practical difference may be seen by comparing what is called the English School of philosophy with the Continental. From Bacon downwards
all our philosophical writers with but few exceptions
(and even in these the theologian has usually pre-
ponderated over the philosopher, as in Berkeley and
Coleridge)—all our philosophical writers are domi-
nated by the notion of a separation between consciousness and its objects, and approach philosophical
questions with the notion of settling what we can
know of objects, with what certainty we can know it,
and what our wisest course of action is in consequence. But this is to adopt the distinction between
the mind in its organism and the world external to
the mind, as an ultimate one. Our English writers
are thus psychologists in the above explained sense

of the term, and not philosophers in the strict sense. All our great triumphs have been won on this basis. Bacon's "Homo naturæ minister et interpres tantum facit et intelligit, quantum de naturæ ordine re vel mente observaverit, nec amplius scit aut potest," shows this in the most unequivocal manner; and so also does the whole First Book of the *Novum Organum*, with its demolition of the Four Idols, and its methods of sound philosophising. The same presupposition is obvious in Locke's[1] disproof of Innate Ideas. Berkeley's Idealism again, based on his Theory of Vision, is a psychological theory; it resolves the connection between consciousness and its material and external objects, assumed as a causal one, into a causal connection between the mind and its states of consciousness. Hume's system, based upon Berkeley's, and applying his principles, evaporated substantial Mind as completely as Berkeley had evaporated substantial Matter. It was the suicide of a non-philosophical psychology, and was immediately followed by Kant's philosophical reconstruction. Hartley is a thorough-going physiological psychologist, establishing the complete dependence of consciousness on its organism. Very rarely does John Stuart Mill rise fairly and indisputably into the philosophical region, and when there he takes but a short flight; one such occasion is when, in his *Examination of Sir William Hamilton's Philosophy*, he defines Matter as a "permanent possibility of sensation."[2] The existence of Mind, to

[1] See Mr. T. H. Green's masterly disquisition on Locke, Berkeley, and Hume, in the General Introduction to his and Mr. Grose's recent edition of Hume's *Philosophical Works*. The truth of what I here state about these writers cannot be more fully or more conclusively shown than by that disquisition.

[2] Chapter XI. page 198.

which he next proceeds, fairly baffles him. Yet this happy invention of a phrase which will render a philosophical conception familiar to English readers is a great service to philosophy. When we come to living writers we find in the speculations of most of them no difference, I at least can find none, in respect to the principle now in question, from those of the great English writers who have preceded them. Take for example Mr. Spencer. Although he distinguishes subjective psychology from objective, and maintains of the former that "under its subjective aspect, Psychology is a totally unique science, independent of, and antithetically opposed to, all other sciences whatever. The thoughts and feelings which constitute a consciousness, and are absolutely inaccessible to any but the possessor of that consciousness, form an existence that has no place among the existences with which the rest of the sciences deal;"[1] and though this might seem amply sufficient as an admission of the philosophical principle of the necessity of a subjective and analytic method; yet Mr. Spencer immediately, and even in this very enunciation, falls back into the separation between the objective and subjective aspects, "Mind still continues to us a something without any kinship to other things;"[2] and Psychology consists of two totally-independent aspects, objective and subjective,—" the two forming together a double science which, as a whole, is quite *sui generis*."[3] Mr. Spencer has not seen that it is Reflection, in subjective psychology, which perceives the two aspects subjective and objective, and that the two, as so perceived, are insepar-

[1] *Principles of Psychology*, Vol. I. § 56, p. 140. 2nd edit.
[2] Id. id. p. 140.   [3] Id. id. p. 141.

able and co-extensive. He speaks of several classes of nervous changes which "have objective aspects only—do not present inner faces to consciousness; and others have subjective aspects in early life, but cease to have them in adult life."[1] If so, I would ask, if these nervous changes have *no* subjective aspect, how is it that he is aware of their existence? Mr. Spencer takes the proximate *conditions* of subjective states (conditions *existendi*) for the objective *aspects* of those states.

Mr. Spencer's conception of the subjective aspect of Psychology, then, would be totally inadequate to serve as a Philosophy, if any one should put it forward to do so; for it is deficient in generality. Mr. Spencer *distinguishes* it from objective science, and this, so far as it goes, would enable it to serve as a philosophy; but he does more, he *separates* it from objective science in separating it from the objective aspects of things. But if there are objective aspects of things which have *no* subjective aspects, as the last-quoted passage shows him to maintain, then the subjective aspect of things, and the subjective analysis which deals with them, must be quite inadequate to deal with things in their most general relations and laws, that is, to philosophise about them.

The comparative narrowness of this point of view is seen when we turn to the development of philosophy in countries where the distinction between man's knowledge and the world external to man was not the dominant one. Beginning with the publication of Telesio's work[1] *De Natura Rerum juxta propria principia* in 1565, we find in Italy a philosophy of

---

[1] *Principles of Psychology,* Vol. I. § 43. p. 104. 2nd edit.

[2] See Signor Fiorentino's *Bernardino Telesio* : Firenze, 1874.

nature ripening into the large all-embracing systems
of Giordano Bruno and Campanella. " Bruno and
Campanella worked at a metaphysic entirely new,
which was to be a metaphysic of identity, to replace
the metaphysic of Aristotle, which may be called one
of duality and opposition."[1] And if these two great
minds still sought to explain the universe by means
of entities imagined out of abstractions, this was no
more than was inevitable for men. in that early age,
who refused to envisage the problem before them in
anything short of its true and vast proportions, and
who would have scorned to claim for themselves the
title of philosophers while leaving others to solve its
hardest questions and encounter its deadliest enemies.
English philosophers on the other hand, in declining
the pursuit of " formal and final causes" as " barren
virgins consecrated to God," were in truth declining
for themselves the arduous attempt to include Theology in the philosophical domain. and were thus
compelled either to accept it ready made from the
theologians, or leave it to be criticised and combated
by others. It is the poets and not the philosophers,
it is our Marlowes, our Shaksperes, our Miltons, our
Shelleys, who in England have been the real antagonists of a narrow and unphilosophical theology.

Three words are the imperishable contribution of
Descartes to modern philosophy, possibly his only in-
controvertible one. But these three words are the
morning-star which ushers in the new day. In the
famous *Cogito ergo sum* is expressed the distinction
between consciousness and its objects, in contrast to
that between man's knowledge and the world external
to man; the fundamental distinction of philosophy

[1] Fiorentino : Work cited, Vol. II. p. 183.

as opposed to the fundamental distinction of psychology; the assertion of the moment of self-consciousness or reflection as opposed to the moment of direct consciousness or direct perception.

In Leibniz, with whose mind all modern Germany is impregnated, we have again a system of philosophy including psychology within it. The monad of Leibniz was not the monad of Bruno; but, says Sig. Fiorentino, "for all that, Bruno and Leibniz have as much resemblance as was possible for two philosophers between whom Descartes had intervened. Bruno, prior to the Cartesian reform, would find the union of opposites in nature; Leibniz, who came after it, in subjective thought, in that power of reflecting the universe which each monad carries within it."[1]

But it was not until Kant that the Cartesian moment of self-consciousness was to become explicit, militant, and finally dominant. And this is the reason of the supreme importance of Kant in philosophy. The difference between the two principles of psychology and philosophy, I mean the two moments of direct and reflective consciousness, the latter involving a distinction without a separation, the former a separation following on a distinction, was dormant and unperceived until Kant, who himself held them both without perceiving their incompatibility, endeavoured to combine them, in the *Critic of Pure Reason.* Kant endeavoured to hold together, as principles equally dominant, the notion of a Mind endowed with faculties and that of a moment of self-consciousness, the so-called unity of Apperception. And Kant's system exploded into fragments because it contained these two principles in this close juxta-

---

[1] Work cited, Vol. II. p. 105.

position. This showed that one of them was a fiction; there are no such things as innate forms either of the faculty of intuition or of the faculty of thought. This is not yet understood by us English. We are still occupied in expounding Kantianism, as if it was a living system. You might sooner rebuild Solomon's Temple. It is just an instance of what I said at the beginning about the use of systems; Kant's system was the means of verifying the principles which he believed himself to have discovered, and resulted in the establishment of some, the discrediting of others.

To have, then, two fundamental principles at once, essentially different yet professing to cover the same ground, is impossible; either one must be retained and the other discarded, or else a *modus vivendi* must be found and a separate function assigned to each. Most of the German post-Kantian systems have attempted to discard the psychological, the English the philosophical, principle; and to select and discard exclusively either the one or the other would be easy enough, if only facts would allow you to ignore the one which you have discarded. But this is impossible. Sooner or later an exclusive philosophy is wrecked on the rocks of science; and an exclusive psychology on the rocks of philosophy. To discover a *modus vivendi* between the two principles, then, and thus to form a single philosophical whole, with its two doctrines, philosophical and psychological, contradistinguished and yet combined, so that each may illumine, control, and support the other,—this has been and is the problem of philosophy from Kant's time to ours. It is not a question of sacrificing either, but of combining both in the places and with the functions which each is suited to perform and fill. Now the philosophical principle has been shown above to be the broader and more general of the two, and the questions which spring from it remain to be answered when the psychological principle has adduced its last proof and said its last word. It is a power which must be reckoned with, since it cannot be either ignored or transcended. And we possess in philosophical analysis a mode of criticising all non-scientific speculations, to the irruptions of which the territory of science is constantly exposed, and against which science has no weapon of its own but that of attempting to ignore them. Philosophy in short is alone competent to deal with speculations which, whether they are tenable or whether they are absurd, spring at any rate from a reflective source, and consequently are of a philosophical character.

The English school of thought was based on the acceptance of the scholastic doctrine of Nominalism as a sufficient basis of philosophy; Nominalism meaning, for the members of that school, the doctrine, that general terms were nothing else but abbreviated *expressions* for the classes of similar particular percepts, or things, of which experience consisted. In this they showed their practical and non-theoretical tendency; being satisfied with a very meagre account, to say the least of it, of what general terms really *were*, and asking principally what they were *good for*, what purpose they served. But whatever may be the truth with regard to the scholastic doctrine of Nominalism, it is totally unfit to serve as a basis of philosophy. Both in its first shape, in which it is identical with Conceptualism, and still more in its later shape, in which it is opposed to Conceptualism and is then more properly to be called *Terminism*,— it leaves behind it a large field of questions untouched, relating to Percepts. When Locke, for instance, maintained *Nihil in intellectu quod non prius in sensu*, another answer than that of Leibniz—*nisi ipse intellectus*—was at hand; an answer consisting in the further question—But what is *in sensu?* This answer it was which was formulated by Kant, a formulation which was itself open to the objection made above, namely, that it assumed the Mind as separable from the World, by assigning one element of sensation to the Subject and another to the Object. Further analysis than Kant's, but in the same general direction, was therefore a necessity. But the English school ignored the possibility of such an analysis of Locke's *sensus*. They assumed sensations as the atoms, so to speak, of consciousness; and even now, though admitting that these atoms may have distinguishable elements, they do not in practice lay any stress on distinguishing them.

Mr. Spencer's *Principles of Psychology*[1] will again furnish us with an illustration. Accepting as really simple those constituents of Mind which are not decomposable by introspection, he mentions two kinds of proximate components of Mind,—Feelings and the Relations between feelings. "Each feeling, as we here define it, is any portion of consciousness which occupies a place sufficiently large to give it a perceivable individuality; which has its individuality marked off from adjacent portions of consciousness by qualitative contrasts; and which, when introspectively contemplated, appears to be homogeneous. These are the essentials. * * * And obviously

[1] Vol. I. § 65. pp. 163-165. 2nd edit.

if it does not occupy in consciousness an appreciable
area, or an appreciable duration, it cannot be known
as a feeling." "A feeling proper is either made up
of like parts that occupy time, or it is made up of
like parts that occupy space, or both."

Here we have what I have called above the pure
time and pure space elements of percepts. The feelings proper, though not decomposable by introspection, have yet elements which introspection
distinguishes. They are empirical units, but metaphysical concretes. Mr. Spencer however leaves behind him their analysis, and passes to the examination
of their interconnection. He goes no farther back
in analysis than is requisite for his Psychology, no
farther than to those units of consciousness which
correspond to his ultimate units of physiology, his
single hypothetical "nervous shocks." But these
units of consciousness are not simple but complex,
if we look at them subjectively. In the analysis of
direct perception, therefore, there is a great field left
untrodden by one of the ablest of modern psychologists.

But the ultimate analysis of perception in reflective or self consciousness, and not merely in direct,
is the question on which philosophical controversy
must chiefly hinge, at least for the present. It involves the question of the possibility of the alleged
"Intellectual Intuition," of envisaging a substratum
common to the two modes of existence, consciousness
and objects of consciousness, and of all the various
forms which this mode of speculation may assume.
And the analysis of direct perception to its furthest
limits, not stopping short at Mr. Spencer's admission
that it can be analysed, is a pre-requisite for the ana-

lysis of reflective. It will not suffice for psychology to throw the *onus probandi*, e. g., the proof that we have a "faculty" of Intellectual Intuition, on supporters of the systems of speculation contemplated. The question is one concerning the *contents* of experience, not concerning its *conditions*. It will not do to say,—we have no "organ" for procuring us such and such experiences; we must first enquire what experiences we actually have, and then will follow the question, what "organs" are those by which they are procured. So long as psychological schools can be fairly taxed with narrowness of basis, with not embracing philosophical problems in all their length and breadth, they may hold their ground as science, but they cannot be regarded as judges in matter of philosophy, or pleaders in matters of theology.

But here I must turn to another point. If the ultimate analysis of perception in all its branches is the distinguishing mark of philosophy,—if this is what its differentia from psychology consists in,—then I shall be asked, Is not philosophy, in your view of it, just what the Germans mean by Cognitiontheory (*Erkenntnisstheorie*), nothing more, nothing less? The answer to this question will, I hope, clear up any remaining obscurity which may possibly be still left hanging about my conception of philosophy. The answer depends upon bringing into connection, and confronting one with the other, what has now been said on the relation of philosophy to psychology, and what was said in the former section of the Chapter on its relation to the other sciences, and especially to the notion of Existence.

The German Cognitiontheory is a child of Kantianism, and it has not outgrown its parent. Philosophy, in my view of it, is more than Cognition-theory, inasmuch as it contains in the phenomena of Reflection, round which it centers, the answer to the question, What existence itself is. It is not only a theory of knowledge, but a theory of existence also.

Under all doctrines, opinions, theories, there lie, perhaps tacitly, *assumptions;* and these are usually to be traced to, or at least when traced are found in the form of, *distinctions.* People distinguish *this* from *that*, and then put the question, Do you mean this *or* that? assuming that you must mean either one or else the other. Their assumption is a distinction which appears to them undeniable. Now Cognitiontheory proceeds on the distinction between knowledge and existence assumed *as an ultimate one;* and thereupon a cognition-principle (*Erkenntniss-princip*) is opposed to a real-principle (*Realprincip*). Thus, (by way of example), Herbart reproaches Fichte with "the fallacy of transforming the Ego, which he was justified in using only as cognition-principle, into a real-principle."[1]

Now there is a great deal of truth in this distinction, but there is also so much error, or to say the least indistinctness, that it cannot be used as the basis of philosophy; that is to say, the German Cognitiontheory, which is founded on it, is not philosophy in its length and breadth. We want to know first what a *Real-principle* is, and what constitutes its distinction from a *Cognition-principle;* we want first to know, as Mr. Matthew Arnold has well urged, and to use his own words, "we want first to know what *being* is."[2] The distinction can be ultimate only if

[1] Allgemeine Metaphysik, § 98. Werke, Vol. III. p. 278.
[2] God and the Bible, Ch. II. p. 72.

each member of it is *per se notum*; i. e., immediately known and incapable of further explanation.

Now my contention is, that this is not the case; but that the meaning of both terms is given only in and by reflection. Thus the facts of reflection overrule the distinction between knowledge and existence, cognition-principle and real-principle, as an ultimate one; those facts give rise to it, and they alone can interpret and explain it. The important fact is, that in reflection knowledge and existence are united, not separated. The term *cognitiontheory* assumes that they may be separated and treated apart. The *true* cognitiontheory is a reflection theory; that is the step forward which philosophy has made since Kant, and chiefly through him. Greater definiteness and deeper insight into cognitiontheory is attained; reflection is its principle, and the facts of reflection *nostri farrago libelli.*

The nature, analysis, or whatness of a thing is a real-principle, to use the German phrase, as much as a cognition-principle; and this is given in reflection; *esse* is *percipi.* There is a double-aspected source of knowledge and of fact, a common source to the two streams, as it were, of investigation and the history investigated. Knowledge and fact are not separate *ab initio.* It is of course important to keep them separate in scientific investigations. It is in philosophy that their original union is apparent, and it is the business of philosophy to insist upon this.

The facts of reflection, then, overruling the distinction of cognition-principle and real-principle as an ultimate one, replace it by another, or rather remove it to a subordinate position, as belonging to history and genesis, not to nature and analysis. *Conditiones cognoscendi, conditiones existendi,*—that is the shape in which the older forms now appear.

The nature, analysis, or whatness of anything is double, subjective and objective. This belongs to reflection and to philosophy. The genesis or history of anything is, not double, but of two separable kinds; there is the history of the thing itself, and the history of our knowledge of the thing; its conditions of existing, and the conditions of our knowing it to exist. And this distinction between *Nature* and *History* is the link between the philosophical analysis of nature, given by reflection, and the scientific enquiries into (1) the conditions of things and events existing and occurring, and into (2) the conditions of our having knowledge of things and events, the *ordo ac connexio idearum*, which is psychology.

Metaphysic (which is the analytic part of philosophy) and psychology have, in one sense, the same object-matter, states of consciousness, but treat it in quite different ways. Psychology is a science, subordinate to metaphysic which is philosophy. Herbart's name must never be forgotten, when this doctrine is in question. The particular way, indeed, in which he drew the line between them, I cannot but think erroneous. But the relation in which they stand to each other, as a more and a less general method of treating the same phenomena, he was the first to dwell on, explain, and enforce. "*The whole series of Experience-forms must be investigated twice over (doppelt); metaphysically and psychologically.* Both investigations must lie side by side, and be compared together long enough for every one to see their complete difference so plainly as never to be in danger of confusing them again."[1] Psychology has to do (in my opinion) with the genesis and history of states of consciousness in particular organisms; metaphysic has to do with the analysis of states of consciousness in connection with their objects; the objective aspect, as a whole, being summed up in the word Existence.

The meaning of the term *existence* is at the root of all philosophical discussion; and more prominently so since Kant. But precisely here it was that Kant was defective. He did not see that *existence* was a *reflective* perception; he saw that it was not an inherent quality or attribute; but he thought that the notion of anything's *existing* was a *positing*, a *Setzung*, of the mind. "*Existence* (*Sein*) is plainly no real predicate, i. e., conception of something which can come on additionally, to the conception of a thing. It is merely the positing (*Position*) of a thing or of certain determinations by themselves."[2]

This notion of existence being a *positing*, which does not necessarily exclude, but at the same time does not distinctly admit, reflective perception,—this notion was taken up by Fichte, and its vagueness determined in the Idealistic sense. The Ego was made to posit, that is, to *produce*, what it really *perceived*. Positing is an action pure and simple; reflection also is an action, but the action is perception;—an object is involved *equally independent* with the Subject. This balance of *equal* independence, *mutual* interdependence, was destroyed by Fichte. *Existence* fell wholly on the side of the Subject, the Ego, which was pure *agency*.

[1] Allgem. Metaph. Werke, Vol. III. p. 258.
[2] Kritik d. R. V. p. 439 : Hartenstein, 1853. From the famous *hundred dollars* passage.

This shows us, too, how the most opposite theories come out of Kantianism. Take Kant's *positing*, and make the balance incline to the objective side, the *thing posited*, and you have an objective existent,— something *there* to be perceived. The fact *that* the thing is there, apart from *what* it is, is Herbart's interpretation of Kant's *positing;* upon this he bases his philosophy; and for this reason only he calls himself a Kantian.[1] But to take this alternative is to take, along with it, what Kant was quitting, what lay behind him, what the philosophical world, before Kant, was familiar with, namely, the supposed absolute character of objective existence.

But it is *in the balance* itself that, not only the relativity of existence, but also its true explanation and analysis, consist. Let it once fall on either side, and two things are the result; first, you have an Absolute Existence, whether absolute as subject or absolute as object, and secondly, you have the question, *what* their existence means, to solve over again; that is, you have to rediscover the relativity, the $esse = percipi$, of existence. In other words, the point of view of reflective perception is the only final one, however unlike its solution may be to what we expected, in answer to the question which we imagined ourselves to be asking. The answer which we get from reflection not only is, but also shows why it is, the final one.

The Germans, then, have not carried out their cognitiontheory to its full limits, have not given it its legitimate development. Yet how forcibly are we again struck with the contrast which their treat-

_____
[1] Allgem. Metaph. Werke, Vol. III. pp. 65. 117. 157. And many other passages.

ment of the question presents to that which is usual in this country. They treat philosophical questions as philosophical; they have not that profound horror of the theoretical, which is the too common characteristic of Englishmen. And this leads me to remark, in conclusion of the present section of the subject, the very different reaction which the larger and the narrower methods of philosophising respectively exercise on their disciples.

The practical result of the larger view of the scope of philosophy, and of the discussions raised by the introduction of the Cartesian moment of self-consciousness into philosophy, both before and after Kant, has been to render philosophy more searchingly analytic. Hardly any analysis of conscious states pure and simple is to be found in English writers; whose strength is expended either on the physiological and physical conditions of those states, or on their sequences in consciousness itself under the title of laws of Association. In German writers, on the other hand, analysis of this kind is very frequent and very excellent. They map the country before exploring it in detail. The works of Leibniz, and especially of Wolff, are storehouses of distinctions; Kant analyses and analyses again, first from one point of view, then from another, making each new analysis throw light on former ones. If of Hegel's great system not one stone should remain upon another, his all-penetrating all-comprehending analyses will for ever remain as instructive and as stimulating to the mental powers, as are those of Plato and those of Aristotle, whose systems have long ceased to find disciples.

Far be it from me to depreciate the powers or the

achievements of my countrymen. I glory in Bentham, in Locke, in Hobbes, in Bacon; I glory in William of Ockham and all his train;

> " πλατεῖαι πάντοθεν λογίοισιν ἐντὶ πρόσοδοι
> νᾶσον εὐκλέα τάνδε κοσμεῖν."

But yet is there a more excellent way; and we shall not merge our individuality by forming an alliance, nor need we strike our colours in setting sail upon a broader stream. *A greater and more comprehensive philosophy can arise in the line of Locke than can ever arise in the line of Leibniz;* but only on the condition of replacing our narrow psychological horizon by an horizon of true philosophical range. This being done, our psychological and scientific method is at least as necessary to the soundness, as the philosophical to the comprehensiveness, of a complete philosophy.

Briefly, then, to resume the position at which we have now arrived, we may define Philosophy, in contradistinction to Psychological Science, as the ultimate analysis of states of consciousness in connection with their objective aspects, abstracting from their conditions in the organism; and in contradistinction to Science in general, as the subjective analysis of the ultimate notions of the Sciences. In both alike it has the three characteristics of being ultimate, subjective, and analytic. The first characteristic, *ultimate*, belongs to philosophy *ex hypothesi*. That is to say, only such enquiries as are ultimate, which stand nearest to and endeavour to penetrate farther into the unknown, the "dark foundations" of being, do we set apart as search and not as science. The second, *subjective*, rests on a simple fact of experience, the apparent reduplication of objects in subjectivity; consciousness being like light, which reveals itself and the object at once; the object itself and the object seen being one. The third, *analytic*, is determined by the process of Reflection being made the principle of the method pursued. But this third characteristic is open to the doubt, whether it entirely exhausts the possibilities of philosophy; whether it does not restrict philosophy to too narrow a field; whether philosophy itself may not be synthetic also. It is clear that philosophy, being subjective and ultimate, must be reflective, and therefore analytic of its object-matter; the question is, whether it is analytic only.

§ 3. The question just raised, whether philosophy is more than merely analytical, brings us for the first time into serious contact with the claims and pretensions of Ontology. The ontological question may be variously stated. Is it possible to transcend the distinction between the subjective and objective aspects, resolving them into something which is neither of them actually, and yet which is both of them potentially? Or again, Is it possible to exhibit the genesis of the two aspects, if not from a common source, yet of either of them from the other? Or again, Can we hope to assign a reason why there should be consciousness and existence at all, or why there should be consciousness and existence only?

These are statements of what is properly and strictly the problem of Ontology. An ontological system is one which professes to furnish and demon-

strate an affirmative answer to any one of these questions. But if we look at the various systems of
philosophy which are synthetic and constructive as
well as analytic, we shall see that they envisage these
questions with very varying degrees of distinctness;
and that, even where they envisage the questions
distinctly, they yet contain much which, though professedly constructive and not analytic, still does not
go beyond the ultimate duality of subjective and
objective aspects, distinct but inseparable, given by
reflection. Let us enumerate the explanations, the
ultimate constructive principles, given by some of
these systems. Such constructive principles are:

Book I.
Ch. I.

§ 3.
The Construc-
tive Branch of
Philosophy.

The Very One, τὸ αὐτόεν, of Neo-Platonism,
The Triune God, of Christian Theology,
Leibniz' Monad of Monads,
Spinoza's Infinite Substance consisting of In-
finite Attributes,
Schelling's Absolute Identity of Subject and
Object,
Hegel's Identification of Contradictories,
Schopenhauer's Will as Thing-in-itself,
Mr. Herbert Spencer's Unknowable.

These are all produced as solutions in constructive
philosophical speculations, widely different from each
other as they are in other respects; for instance, in
respect of how much, in detail, the several systems
they belong to profess to have achieved; or again,
from a theological point of view. One and all, they
go to the very limit of the human tether; they embrace *everything*. Even Mr. Spencer's system, which
knows that nothing can be known of the dark foundations, is unlimited knowledge in respect of range. It

Book I.
Ch. I.

§ 3.
The Construc-
tive Branch of
Philosophy.

knows so far, and *all* beyond is known to be unknow-able. Hegel's system, on the other hand, knows so far, and knows that *all* beyond is knowable too, for its inmost source of being is known.

It is very difficult to pronounce whether the authors of systems such as these suppose themselves to have transcended the duality of subject and object, for they imagine something as transcending it, and by this very act of imagining make subjective that transcendent something. The truth is, that they have not grasped, or at least have not stated, in all its bearings, this property of Reflection, namely, that *all* its objects are subjective as well as objective; either they have not repeated often enough the process of reflecting again on their own proceeding, or else they have not thought it worth while to explain the *rationale* of it to their readers. At the same time they all of them go, as remarked above, to the very end of the matter.

We see, then, two things; (1) Constructive systems have busied mankind in all ages; (2) the ontological part has not been clearly marked out from the re-maining parts of those systems, and thus Ontology has not been distinguished from a possibly legitimate philosophical construction.

The necessity for leaving a place open for philo-sophical construction, distinguished from Ontology, is shown by the facts of Reflection. Consciousness (I mean always as it exists in an individual) stands as it were at a point in space, an infinite space, which is Existence. But consciousness is as infinite as the space. Otherwise, how could it imagine the space as infinite? Again, consciousness is endowed with facul-ties as it is called, that is, works in certain determi-

Book I.
Ch. I.
—
§ 3.
The Construc-
tive Branch of
Philosophy.

nate modes; now everything determinate is limited by its determination,—there may be other modes which are to this consciousness indeterminate, and in that sense unknown. The world as given by our determinate modes of consciousness is to us the actual world in which we live. The world as it would be given by other modes is indeterminate existence, indeterminate to us but determinate to those other modes of consciousness, being their actual world. There is then, beside our determinate world, a world indeterminate to us, but *possible* if there should be other modes of consciousness than ours, that is, *possible* to our thought since we imagine its condition, and *actual* to those other modes, if *they* are actually existing. It is this possible and *to us* indeterminate world which is the field of constructive philosophy as distinguished from analytic philosophy on the one hand and from Ontology on the other. It is a world included within the grasp of Reflection, but excluded from that of Direct Consciousness.

The constructive branch of philosophy, then, having this field assigned to it by the method of Reflection, is not only one which has been pursued by philosophers, but it is a necessary and legitimate branch of enquiry. The direction in which its problems and their solutions (if any) will lie, and the point of its connection with the analytic branch, are evident. How much it will contain, what answers are possible to its questions, and even the particular shape of these questions, are points which need not here be dwelt on. One thing however is too important to be passed over. The constructive branch of philosophy cannot be pursued except in connection with the analytic branch. Otherwise it would

Book I.
Ch. I.

§ 3.
The Construc-
tive Branch of
Philosophy.
not be philosophy at all,—for it would not be subjective and it would not be ultimate. But it does
not necessarily follow that its problems must take an
ontological shape, that philosophers should propose to
transcend the distinction of subjective and objective
aspects, and to define the supposed underlying Unity.
We have got two analyses, a subjective and an objective, to combine. The elements given by these
analyses may be hypothetically constructed and recon-
structed in various ways. Again, the modes of consciousness known to us are not all that are possible;
there may be conscious beings with senses, emotions,
and intellectual endowments, widely different from,
as well as more numerous and more powerful than,
our own. There is a vast field of the unknown to
traverse, and all this field lies, with all its possibilities, before the constructive branch of philosophy.
Science is of the world as our human modes of consciousness reveal it to us; it is a knowledge of the
laws of objective existence as that existence appears
to man : it is the history of the universe expressed
in general terms, in concepts not in percepts. Still
it does not transcend what has or what may have
happened, what can happen or will happen, as conceivable by ourselves. But a Fourth Dimension of
metaphysical or unfigured Space is a problem of constructive philosophy. It is an object not unknowable, but unknown; we know the *sort* of thing it
must be, at the same time as we know that we cannot
construe the notion of it to ourselves, or bring it into
harmony with our actual space of three dimensions,
which seems to include all possible directions of
spatial extension.

Or we may state the position of the constructive

Book I.
Cii. 1.

§ 3.
The Construc-
tive Branch of
Philosophy.

branch of philosophy as follows. There cannot be anything, beyond existence, that is not existence; it is a contradiction in terms ; or, there cannot be an existence that has not a subjective aspect, for to *be* is to have a subjective aspect, in some consciousness or other. But there may be existences, or existent worlds, very different from that one which is the world of our consciousness ; and we may imagine such worlds, analogous to our own, by supposing changes in regard to the ultimate elements of our own subjective analysis, which may be done either by the hypothesis of different but analogous Feelings, or by that of a different combination of the subjective elements themselves. There would thus be new worlds for new conscious beings. And our own world would appear as one of a series or *group* of analogous worlds, yet without in any way transcending phenomenal existence, or the duality of subjective and objective aspects, distinct but inseparable. We may farther suppose this series or group graduated so as to consist of members infinitesimally different from each other, yet none of which would be actual to any conscious beings but its own, however near it stood to those most like it in the group. And this whole hypothetical group of phenomenal worlds would be the field of the constructive branch of philosophy.

Again we may say, that the problems of the constructive branch of philosophy will include, but possibly by giving them quite different shapes from what they have hitherto appeared under, the old questions of God, and Immortality; and generally of the *Why*, the *Whence*, and the *Whither*, as distinguished from the *What* and the *How*, of our actual world. I exclude

Book I.
Ch. I.

§ 3.
The Construc-
tive Branch of
Philosophy.

the question of Freedom, that appearing to me to be not a problem, but a puzzle which is within the competence of the analytic branch; though it is not impossible that further light should be thrown upon it from the constructive. But if we take our actual world as a whole, limited by the worlds of the hypothetical group, then the *How* of our actual world becomes a question of its connection with those worlds, and consequently a question for the constructive branch of philosophy.

Whatever light might be thrown upon the nature of our own world by imaginations directed to the hypothetical group, the insight gained would not be science; no laws of our own world would be thereby discovered or proved; for the hypotheses would not be verifiable. Still there might be an advantage to science in suggestions which might be thus afforded for framing other scientific hypotheses relating to our own world, hypotheses which themselves, when framed, would be capable of verification and therefore scientific. For instance, but by way of illustration merely, the hypothesis of an universal Ether is a scientific hypothesis, capable of verification by means of its consequences; but it is also an hypothesis which we may easily imagine to have been suggested, in the first instance, by speculations on hypothetical worlds or their relations to our own. And supposing that to have been the case, we should then have an instance of what I intend by saying that hypotheses in philosophy, not themselves verifiable, may conceivably suggest hypotheses in science which are or may become so. Here is the link which connects the constructive branch of philosophy with science in its strict signification; and it is clear I think that, though indirect, it is still not to be altogether disregarded, even from a scientific point of view.

Widely different however from what is here intended as the constructive branch of philosophy are theories which bear a similar relation to some particular branch of science; such theories as for instance that of Herr G. Th. Fechner in his *Zendavesta* (1851), to the effect that the whole universe is animated, and every sun and planet a being endowed with an individual life and consciousness; or that put forward in *The Unseen Universe* (1875), which maintains the existence of an infinite series of worlds composed of finer matter than that out of which the visible universe is framed. One is a physiologist's, the other a physicist's, imagination. Such theories belong to a constructive branch, not of philosophy, but of science. They build an extension, or they remodel the structure, of the physical world as we see it and know it; but they build or remodel with the old stones, or with new ones of similiar quality. Philosophy in remodelling would transmute the very substance of the stones as it were into diamond or opal. I am very far from denying the value of such scientific speculations; I welcome and applaud them. At the same time it is necessary to keep them distinct from analogous imaginations of philosophy in its constructive branch; and, above all, not to use them as supports of a ready-made theology, which is thereby made to sit in the chair of philosophy.

It is at this point that I find myself opposed to those who hold the fourth of the views concerning philosophy enumerated at the outset, a view recently and very powerfully advocated by Mr. Lewes. The field which I have vindicated as belonging to a legiti-

84 — PHILOSOPHY AND SCIENCE.

mate branch of philosophy, the constructive branch, is, if I mistake not, that described by Mr. Lewes as the region of "the supra-sensible" and "the metempirical," and excluded, in the character of an "unexplored remainder," not only from science but also from philosophy.[1] At the last of the places cited, Mr. Lewes mentions two senses of the word Object. "We apply the term Object to the Not-self. This Not-self may be either the objective aspect of the world felt and thought, i.e., of the External in actual and virtual relation to Sentience; or the universe of existence, conceived in its totality, including that smaller section of it which is grouped by a Subject." And then, speaking of the "Universe considered as the totality of Existence," under this aspect he says, "the Object is not the *other side* of the Subject, but the *larger circle* which includes it."

On this I would ask,—This interspace between the including and included circles,—what is it? Is it knowable or unknowable? If unknowable, how do we know that it exists? If knowable at all, why not included in philosophy?

There is one way of understanding the expression "larger circle," in which it does not conflict with the expression "other side," here contrasted with it. Namely, when the universe considered as the totality of existence is distinguished into its two aspects, the Object and the Subject, this double-aspected Whole may be taken as the *larger circle* including either of the two aspects, since it includes both Object and Subject; but then this same double-aspected Whole is also, in the present act of reflection, the Object of

---

[1] Problems of Life and Mind, Vol. I. pp. 39-46. 177. 189. 194-5.

our imagination; it is the *other side* of our subject.
Reflection continually distinguishes, equates, and
combines, the two aspects. And in this sense the
universe, the totality of existence, is the object of
philosophy, being the object of Reflection. It is
from the point of view of direct consciousness that
the universe is *not* the other side of the Subject, *but*
the larger circle which includes it, since it includes,
in Mr. Lewes' words, "that smaller section of it which
is grouped by a Subject."

There is then, according to Mr. Lewes, and from
the direct consciousness point of view, an objective
existence which has no subjective counterpart, is not
"grouped by a Subject." This part of existence I
understand Mr. Lewes to exclude from philosophy
under the names " supra-sensible" and " metempirical." " I am far from implying," says Mr. Lewes,
" that a Supra-sensible does not exist. I only affirm
that it does not exist *for us* as an object of positive
knowledge, though forced upon us as a negative conception."[1] He would " divide the field of Speculation
into the Sensible World, the Extra-sensible World,
and the Supra-sensible World: a division corresponding with our previous distribution of positive, speculative, and metempirical."[2] And after speaking at
length of the two first of these, as comprising all
that is accessible to experience, and consequently all
that is admissible in science, he proceeds: " There is,
however, a third division claimed by Theology and
Metempirics, the region of the Supra-sensible, or
Metempirical, which is closed indeed against the Method of Science, but is open to Faith and Intellectual
Intuition."[3]

[1] Vol. I. p. 252.   [2] Vol. I. p. 253.   [3] Vol. I. p. 264.

Book I.
Ch. I.

§ 3.
The Construc-
tive Branch of
Philosophy.

I on the other hand, from the reflection point of view, would include this region in philosophy, though excluding it from science; not because it is "open to faith and intellectual intuition," but because it is embraced by reflection. Unless it were embraced by reflection, I do not see how it could be examined at all, even to be rejected. To examine it at all is to include it in philosophy; the examination may result in its exclusion from science, but not in its exclusion from philosophy, for that would stultify the examination itself. Within philosophy, accordingly, I would assign this region a place as its constructive branch, making this branch of philosophy wholly dependent upon the results of the analytical branch. To include it in philosophy, at least as a possible branch, is to give a meaning to Wolff's definition of philosophy, "*scientia possibilium quatenus esse possunt.*"[1] To exclude it is to leave unattempted the very questions which most torment us, and for the hope of solving which in great measure philosophy itself is undertaken. Indeed for most people, philosophy means the constructive branch, means the solution of the questions, Why, Whence, Whither, of Existence, and means nothing else. The purely analytical branch of philosophy has for most people no interest and no significance.

The exclusion of what Mr. Lewes calls the supra-sensible and metempirical from philosophy, the rejection of a constructive by the side of the analytical branch, is tantamount to reducing the object-matter of philosophy to the dimensions of the object-matter of science; a reduction of field which is in conflict with the larger method of philosophy, the method of

[1] Logica. Disc. Prælim. § 29.

Book I.
Ch. 1.

§ 3.
The Construc-
tive Branch of
Philosophy.

reflection, as compared with that of science. The larger method requires and involves a larger field or object-matter. Mr. Lewes is therefore quite consistent with himself in reducing the method of philosophy to the method of science, as well as reducing the dimensions of its field. "It is towards the transformation of Metaphysics by reduction to the Method of Science that these pages tend."[1] The solution of metaphysical problems by the method of science— this is what I take to be the purpose of philosophy, according to Mr. Lewes; and philosophy with him, so far as it differs from science at all, must mean the analysis which effects the reduction of philosophical to scientific problems, by reducing the method of philosophising to the method of science, and thus prepares their final scientific solution.

This is why I described Mr. Lewes as placing the function distinctive of philosophy from science in the negative task which it performed, of disproving and banishing ontological entities. If this appears an inadequate description, as Mr. Main has urged against me that it is,[2] I think it is only from its unavoidable brevity that it appears so. I certainly had no intention of denying that Mr. Lewes' treatment of the ultimate generalisations of the sciences included "reinterpretation and analysis;" I do not see how they could be treated at all without doing so; nor yet that Mr. Lewes aimed at reducing them to "terms of Feeling."

The one difference, I apprehend, which is the source of the rest, between Mr. Lewes and myself, is

[1] Problems, &c. Vol. I. p. 5. And see the whole of the Introduction *passim*.
[2] In Mind, No. II. p. 292.

Book I.
Ch. 1.

§ 3.
The Construc-
tive Branch of
Philosophy.

no small one. It relates to the method or methods of philosophy and science;—has or has not philosophy a peculiar method, based on the principle of Reflection? The answer to this question determines its distinctive characteristic with respect to science, and the range of object-matter proper to each. The facts of Reflection, as I contend, make the method of philosophy what it is, and inevitably render its object-matter larger than the object-matter of science. Of course I do not deny that reflection enters into the special sciences and into psychology; they could hardly have been constituted without it. A glance at the two former sections of the Chapter will show this. Reflection is the common thread running through all, and connecting them with philosophy. But philosophy elevates this common thread of reflection into a *method;* and it is its method founded on reflection that at once distinguishes philosophy from the sciences and gives it a larger field. Nor do I see, on the opposite view of the method taken by Mr. Lewes, and on the consequent exclusion from philosophy of the supra sensible, of that part of existence outside of "the smaller section of it which is grouped by a Subject," how philosophy can logically advance a claim to be a doctrine concerning existence as a whole, or can make, as Mr. Lewes nevertheless does,[1] any statement about "the Absolute."

But the distinction between the two branches of philosophy now indicated and I hope justified, the analytical and the constructive, is practically most important. It is mainly for want of such a distinction that those of the above-named constructive systems which offered a positive solution have failed to

[1] Problems, &c. Vol. II. p. 503.

Book I.
Ch. I.

§ 3.
The Construc-
tive Branch of
Philosophy.

recommend themselves. For they aimed at giving an explanation of the universe which should contain its analysis and genesis *at once*, in a single principle. They attempted too much, and for that very reason they performed too little. Consider it thus. An analysis of the actual world giving us a cause of possible worlds,—how insufficient must any *such* cause be. How narrow a conception, called in to explain how vast a problem!

We must, then, distinguish two legitimate branches of philosophy, the analytic and the constructive. But the analytic branch has already been provided with a name; it is that which I at least have always spoken of by the name of METAPHYSIC. The constructive branch may remain at present undesignated. The kind of problems which will be attempted in it is as yet too undetermined. Indeed it has never hither-to, so far as I know, been distinguished as a legiti-mate branch of philosophy, as separate from the analytic branch which is Metaphysic, and at the same time cleared and clarified from the pretensions of Ontology, which have been mixed up with it into an undistinguished total. And this is my point of dif-ference from the third of the opinions about philo-sophy, enumerated at starting, that of the Absolutists. Still less has this new constructive branch been fol-lowed with any distinct consciousness of its scope and of the means at its disposal. Nor do I mean, in the present Chapter at least, to make any preten-sions on its behalf, either as to the soundness of the methods possible to it, or as to the results which may be expected from them. All I say is, that theoreti-cally a legitimate place is open for it in the whole of philosophy; that philosophy has such a branch,

Book I.
Ch. 1.

§ 3.
The Construc-
tive Branch of
Philosophy.

clearly distinguishable from Science on the one side and from Metaphysic on the other. And we can see already, in a very general way, what sort of a content it will have. It will consist in the combination of an hypothetical psychology with Metaphysic. It will be hypothetical psychology, psychology carried up into more general regions, because it can only advance by assuming consciousness to be separable from its objects and conditioned by its organism, whatever that organism may be. It cannot, like the analytic branch, *begin* with the objective aspect, but must begin with the subjective, as the only one which is known to it. This it does in making the above-mentioned hypothesis of changes in regard to the ultimate elements of our own subjective analysis. And it ends with the objective aspects corresponding to this new beginning. That is to say, its method is that which we have seen is the method of psychology proper. Its aim is to put the objective aspect, a new hypothetical world, to the hypothetical subjective aspect with which it begins. It is thus closely connected with what is the great problem, as yet totally unsolved,[1] of scientific psychology, namely, What is the mode of connection between consciousness and its organism; or, in what way is it conditioned by its organism? We know at present nothing more than the mere fact of the dependence.

The constructive branch of philosophy is accordingly to be regarded as a philosophised psychology, or the return of Metaphysic upon psychology. The most abstruse problem of psychology is the starting-point of the constructive branch of philosophy. Yet

---

[1] Mr. Lewes takes a wholly different view, which I cannot in this place discuss.

Book I.
Ch. I.

§ 3.
The Construc-
tive Branch of
Philosophy.

the constructive branch is not a higher branch of philosophy than the analytic. This Ontology professed to be. But it is clear that, however large and sweeping we may suppose the constructive branch to be, as it has now been sketched, still it is and must always be impossible for it to transcend the ultimate distinction of subjective and objective aspects, or to resolve these into a higher unity; for this would be to overpass the very limits, to abolish the very distinction, to the establishment of which it owes its own existence as a branch of philosophy. Above and beyond all other branches of knowledge is that subjective method whose last word is ANALYSIS.

I shall have occasion in future chapters to criticise ontological and absolutist systems belonging to the third of my five classes of philosophies. But the second class, Comtian Positivism, which claims to be (1) philosophy, (2) subjective philosophy, and (3) to reject metaphysic altogether, requires a few words of criticism in the present place, before dismissing it from further consideration, along with what I have called English Positivism, as not coming up to the notion of philosophy at all.

"Comte's philosophical position," writes one of the very ablest of his English disciples,[1] "may be summed up in these two sentences :

1. He attempted a Synthesis of scientific conceptions.
2. That Synthesis was *subjective*, and not *objective*.

It discarded, that is to say, all attempts to stand outside the universe, to regard it as a whole, and to

[1] Dr. J. H. Bridges in *Fortnightly Review*, June 1877. Vol. XXI. N.S. p. 853. *Evolution and Positivism*.

Book 1.
Ch. I.

§ 3.
The Construc-
tive Branch of
Philosophy.

explain it. The unifying influence, that which made
it a synthesis, was the recognition of Man as the
central object ; of the study of social and moral phenomena as the central science, to which the rest were
subsidiary."

It will be seen at once that this sense of the term
subjective is not that which I have contended for in
this Chapter; the distinction between *objective* and
subjective not being drawn from the point of view of
Reflection. The recognition of Man as the central
object, and the study of social and moral phenomena
as the central science, do not make this philosophy
subjective in any other sense than psychology is subjective. The arguments which weigh against the
claims of psychology to count as philosophy are
equally weighty against the claims of this philosophy
to do so. *Subjective* means to the Comtist nothing
more than *practical*, or practically advisable; being
central in a philosophy in which everything is referred to the needs and the powers of *Man*.

Of course it is not surprising that, from this practical point of view, metaphysic should be wholly discarded. Still there is an equal misconception as to
what metaphysic is, allied to that as to what subjectivity is, which it is important to notice. For
the Comtist mind will often be found to be most
thoroughly metaphysical in the true and, to me, lau-
datory sense of that term. The term *metaphysical* by
no means suggests to the Comtist *subjective analysis* ;
it denotes the very opposite, namely, the adoption of
unanalysed entities as scientific explanations of phenomena. Thus in the concluding part of Dr. Bridges'
article just quoted,[1] we find him setting down the

two fluids in electricity, phlogiston in chemistry, and caloric in thermology, as unverifiable conjectures evolved by a metaphysical stage of thought. But, right or wrong, warrantable or unwarrantable, conjectures of this kind have nothing *metaphysical* about them in the true sense of the term. *Ontological* their character may be; but it is the very purpose and business of metaphysic to analyse and resolve such entities whenever they come forward in philosophical matter, and of thought akin to metaphysic to do so whenever they come forward in scientific matter. Analysis is the business of metaphysic; and of all its congeners.

One word in conclusion as to the permanent motives for philosophising. Science has its existence and development assured to it by the various utilities which it procures, as well as by the satisfaction which it affords to the deep-rooted passion for pure knowledge. Philosophy has this latter guarantee in common with science, but the utilities which it procures are not so obviously and inevitably manifest. They are nevertheless equally real and equally necessary, that is to say, depend solely on philosophy as much as the others depend solely on science. They belong to the moral more than to the physical world. All the moral sciences, the sciences of Life and Manners, depend upon philosophical analysis in the last resort; the philosophies of Religion, of Morals, of History, of Law, of Æsthetics, seek the definitions, the divisions, and sub-divisions, of their object-matter in the distinctions which general subjective analysis alone supplies. The history of the development of

Book I.
Ch. I.

§ 3.
The Construc-
tive Branch of
Philosophy.

Religion, of Morals, of Civilisation, of Law, of Art, is not enough. Tracing their gradual changes from an earlier to a later mode of existence does not tell us *what* they were at first, nor *what* they are now. At every stage which they traverse or attain, their condition at that stage requires analysis, that we may know what it is that they have been, what it is that they have become. Psychology cannot explain Shakspere, nor analyse the Fourth Gospel. We want another analysis, not to supersede but to complete the psychological. We want an Organon for the *Literæ Humaniores*. The only Organon which is at once sufficiently subtil for this purpose, and sufficiently comprehensive to embrace science as well as literature, is that which is offered by philosophical analysis,—as yet indeed in its infancy, but it is the infancy of a giant. These are motives which must act with increasing force as intelligence expands and strengthens; and on their permanence depends the permanence of philosophy.

There are yet motives of another kind which concur to the maintenance of philosophy. The problems of the constructive branch of philosophy suppose the solutions of some at least of the problems of the analytic branch. This analytic branch, then, which I call Metaphysic, has the key, *if there be a key*, to the questions which concern that larger imagined whole of which the actual world, as science discovers it, is a part, the sea of possibility out of which the island of actuality rises. If there be a key—yes, *if*; but so long as science cannot say there is *no* key, does any one suppose that mankind will cease to look for one? Religion alone forces us to the attempt; for God, restricted to a finite region, cannot be an

Book I.
Ch. I.

§ 3.
The Construc-
tive Branch of
Philosophy.

object of worship. The moral law compels us; for can that law which *conscience* obeys be a law of less than eternal validity?

The problems of the constructive branch, therefore, have an inherent power compelling our attention; they have also an attracting interest; an interest partly of the same kind as the intellectual interest of pure knowledge, and partly also moral. The possibilities which these problems envisage are possibilities of an emotional kind, as well as an intellectual. They have a fascination for meditative minds, a fascination which perpetually induces such minds to frame hypotheses, and calls forth as perpetually the reaction of analysis and criticism. Nor will an hypothesis which is philosophical admit of other than a philosophical criticism. Philosophy alone can understand philosophy; the criticism must be of the household of the hypothesis. The two branches of philosophy, then, which alone are adequate to interpret and explain each other, the one perpetually inventing, the other perpetually analysing and criticising, hypotheses, rest ultimately on the same motives, and spring from the same reflective root. The only competent criticism of the constructive branch is furnished by the analytical; and, so long as there remains anything unknown to be discovered, not science but philosophy itself, that analytic philosophy which is Metaphysic, will remain the speaker of the last word.

CHAPTER II.

§ 1. Mr. Lewes in the two first volumes of his Problems of Life and Mind has performed an eminent practical service to philosophy; he has stepped down from the "irreconcilable" position of ultra Positivism, and has consented to discuss metaphysical questions. Since Mr. Lewes proclaims himself to be still a Positivist,[1] this proceeding on his part is a concession, and a very important one; it concedes the claims of metaphysical problems to be discussed; claims which Comte disallowed, discussions to which he held they had no right. We may argue from this concession being made, that the same validity in metaphysic, which has gained a hearing for its problems, will show itself farther in gaining a recognition

---

[1] "This is not a retreat," he writes, "but a change of front. Throughout my polemic against Metaphysics, the attacks were directed against the irrational Method, as one by which *all* problems whatever must be insoluble." Problems, &c. Vol. I. p. 6. And again, "Nor can there be any dispute that the speculations he [Comte] had in view are inane, when pursued on the Method traditionally followed; but an extension of the principles of Positivism may legitimately include even these speculations; and Scientific Method, rightly interpreted, will find its employment there." p. 7.

for that peculiar principle in its method which, while distinguishing it from science, enables it and it alone to deal satisfactorily with those problems which have thus forced themselves into notice. Mr. Lewes treats metaphysical problems, but treats them, professedly at least, by methods of science ; if he can give satisfactory solutions of them by methods which are strictly scientific, there will be no need for applying metaphysical methods, and all that *distinguishes* metaphysic from science will either vanish entirely, or at any rate become practically useless and therefore pernicious.

This, then, is the position in which metaphysicians are placed by Mr. Lewes' book. They are called upon to face a new aspect of the controversy, not indeed new to their anticipation,—

"non ulla laborum
nova facies inopinave surgit,"

—but new in its actual occurrence, new in the particular shape and dress which it assumes. A scientist professes to solve their problems for them by methods of science. Two questions, then, have to be answered, first, Are the problems really solved ? and secondly, If so, are they solved by strictly scientific and not metaphysical methods? Now any separate doctrine or branch of knowledge is constituted such, and contradistinguished from others, by some peculiarity in its principle and method, not solely by the objects or problems to which it is directed. Its principle and method indeed give to those objects and problems a new shape and colour ; and in this way it may be said that the doctrine is contradistinguished from others by its objects, namely, by its objects as shaped and moulded for enquiry by its principle and method. But in the last resort it is the principle and method, shaping the object-matter, which constitute and individualise the doctrine. Other doctrines will shape and mould those same objects differently, will handle them in their own way, make them new and adopt them thus renewed as their peculiar object-matter. If metaphysic has not a distinct and individualising principle and method, or if that principle and that method are not valid, it cannot sustain itself above or apart from the other sciences, but must leave its problems to be dealt with by them, or not dealt with at all.

Since, then, it is a thing inevitable that I should have frequent occasion to criticise and dissent from Mr. Lewes, let me here express at the outset my great admiration of his two brilliant volumes, as well as my obligation to them for rousing me to examine more minutely the foundations of my own system, and to give renewed attention to its logical structure. Mr. Lewes and Mr. Spencer are the two leading representatives of philosophy in England. It is, then, almost a necessity of the case, that any one who believes himself in possession of a more complete truth should endeavour to make it good, by placing his analysis and interpretation of phenomena in contrast with theirs. But to proceed.

Some few years ago Professor Huxley, in his *Lay Sermons*,[1] held out what seemed to be the hand of fellowship from science to metaphysic; and I attempted to respond to that appeal by showing what the claims of metaphysic were, and what were the lowest necessities in point of principle and method, which metaphysic could not renounce without ceas-

[1] *On Descartes' Discourse* : Lay Sermons, p. 371.

ing to exist, a principle and a method which, being founded in the nature of things, must be recognised by all who make the truth of things their object. I refer to an article which I published in the Contemporary Review for Nov. 1872, on *The Future of Metaphysic*. What Professor Huxley thought of the view there taken of metaphysic I do not know; he has not, I believe, spoken upon the point since that time. But now we have, in Mr. Lewes' book, a distinct attempt to deal with that set of questions which are commonly called metaphysical, and to solve them by methods which are, avowedly at least, not metaphysical but only scientific. The controversy has thus moved a step onwards. There are, so to speak, counter claims put forward on the part of science, to those which I had put forward, in that article of Nov. 1872, on the part of metaphysic.

Now I am not going to review Mr. Lewes' book. I am occupied, not negatively in criticising others, but positively in establishing some of the fundamental doctrines of philosophy. And the reason why I make mention of Mr. Lewes, and the new phase which he has caused the old controversy to assume, is this, in order that the practical bearing of the present work on the immediate philosophical questions of the day may be made apparent. Mr. Lewes has come nearer to the true philosophical position than any other scientific writer. He therefore enables us better than any other to see in what the difference consists ; he has narrowed the limits within which it lies. I shall therefore take leave to advert to his book, whenever the connected course of exposition of my own views leads me to controvert his.

Philosophy was shown, in the preceding Chapter,

to be distinguished from science by being an exercise of reflective as distinguished from direct consciousness. Its principle is the mode of self-consciousness, and its method is prescribed by that principle; it consists in a repeated analysis of phenomena as they are *in* consciousness, as parts or states of it; and not in their character as objects outside consciousness for consciousness to stand and look at. For objects in this latter character, as objects for consciousness to stand and look at, may be treated *also* as objects *in* consciousness; and this way of treating phenomena is therefore the more general, being applicable to *all* phenomena, and not to those only which have the character of objects for consciousness to stand and look at.

This more general treatment is the method of reflection; and reflection or self-consciousness it is which we have now to examine and analyse more thoroughly than could be done in the foregoing introductory Chapter. We have to show its nature and characteristic differences from other modes of consciousness, those other modes having hitherto been included in the one term *direct* consciousness, a term which we shall now find admits a further distinction, namely, into a mode before and a mode after reflection, the three together being those which give the title to the present Chapter,—primary, reflective, and direct consciousness.

In the exposition of the general outlines of philosophy, this distinction of reflection from other modes of consciousness holds the place once held by a general distinction of the "faculties" of the mind, as they used to be called; such as the faculties of sensation, perception, imagination, memory, appetition, volition, reasoning; or rather of the general heads to which these faculties were referred, as for instance, the three groups usually adopted of feeling, cognition, and conation. These are now superseded by the distinction between the two modes of simple (including primary and direct) and reflective consciousness, and their subdivisions; the so-called faculties of the *mind* being exhibited as modes of *consciousness*, and as derivative modes of these two chief and fundamental modes. The distinction, then, between simple and reflective consciousness will prove fertile in other distinctions, all of which will receive their final explanation by being brought into connection with this.

Observe the continuity given to the whole of philosophy by this conception. There is no longer a mind, an immaterial substance, with its several distinct and ultimate faculties, of which no further account can be given, but which, with the mind, are employed to explain the genesis as well as to classify the characteristics of the phenomena of consciousness. But there is a single broad and ever broadening stream of consciousness, the phenomena of which, sensations, emotions, reasonings, volitions, and so on, are given by observation, analysed by experience, and classified under the ultimate distinction of consciousness simple and reflective.

Here too, as so often happens, the poets have anticipated us:

> "The essence of mind's being is the stream of thought,
> Difference of mind's being is difference of the stream;
> Within this single difference may be brought
> The countless differences that are or seem."[1]

[1] From an exquisite poem in the volume entitled *Studies of Sensation and Event*. By Ebenezer Jones. London, 1843.

And there is no question possible concerning the origin of consciousness *as a whole*, or, what is the same thing, of consciousness so far as its *nature* is concerned. This question is possible only concerning a particular consciousness, a part of the whole. And then this question, concerning the origin or conditions of existence of a consciousness, is subordinate to, and a part of, the larger question concerning the nature of consciousness; since any particular consciousness is a part of the subjective aspect of the universe, and any objects which are its conditions of existence are a part of the objective aspect. These two aspects of the universe, however, are inseparably connected by reflection as aspects of each other; and there is no universe other than these two aspects themselves.

We shall find as we proceed, that many vexed questions, which baffle the merely scientific method based on direct consciousness, are readily solved and set at rest by the application of this distinction, that is, by the philosophical method which recognises and applies it. Philosophy alone can solve the problems of philosophy. But reserving all such questions, it is requisite in the first place to show what reflective consciousness is, wherein it differs from the primary and the direct, and why it is a valid basis for philosophy. There is a true and a false, or rather many false ways of regarding reflective consciousness. No one denies its existence; that is not possible. But the interpretation of the facts which constitute it, that is, its analysis and the consequences of that analysis, can be variously assigned. Hence the need of repeated analysis by different persons. Philosophers have usually erred on the side of the too much, scientists on the side of the too little. Ferrier, for instance, to cite the most illustrious name in modern British philosophy, thus begins his *Institutes of Metaphysic:* "The Primary Law or Condition of all Knowledge.—Along with whatever any intelligence knows, it must, as the ground or condition of its knowledge, have some cognizance of *itself*."[1] Here Ferrier errs on the side of the too much. He denies that reflective consciousness is *in any sense* derivative; he derives, on the contrary, all other consciousness from reflective, in the words " as the ground or condition of its knowledge." Besides which, he does not actually draw the distinction between simple and reflective consciousness, but leaves it to be divined and inferred, merely making self-consciousness the ground of all other, and allowing both to appear as united in one act of knowledge. Other criticisms which might be made on this proposition I pass over. But observe, that it is the plain adoption of this principle of self-consciousness, however misconstrued and unwarrantably enlarged, which makes Ferrier a philosopher as distinguished from a scientist.

I come now to another imperfect way of regarding reflection, in one respect similar, in another opposite, to Ferrier's; it is that taken by Mr. Lewes. Mr. Lewes has a hold of the true metaphysical doctrine in this matter; he sees clearly that reflection is derivative; and partially he sees that it does not separate the subjective and objective aspects. But though he adopts the metaphysical distinction, he handles and employs it like a scientist. By which I mean, that he treats the universe, in which the two aspects are discerned, as a thing for consciousness to

[1] Section I. Prop. I.

stand and look at, as what I call an object of *direct*
consciousness: just as if he had taken (what of course
I am not supposing was really the case) my diagram,
given in *Time and Space*[1] to aid comprehension, as if it
was a picture, as if the universe of thought was a ball,
half subjective, half objective. He does not apparently
see, that the act of discerning the two aspects in any
phenomenon is itself the subjective aspect of the phenomenon as a whole; but he treats the phenomenon
as having its subjective aspect in itself as an element, or
as he says a *factor;* thereby retracting the *method* of
distinguishing the aspects, while retaining its result
and treating the object with two factors as an object
of direct and not reflective consciousness.

The matter is mainly dealt with in the First Chapter of the Problem entitled *The Principles of Certitude,*
and there we find him saying, "The conception here
brought forward insists upon the external Real as the
complementary factor of the internal feeling."[2] And
again, "Objective factors (not otherwise to be speci-
fied) existed as permanent possibilities, which might
become Reals when combined with subjective factors."[3] Again, "The world arises in consciousness—
not as the product of the subject only, but as the product of object and subject."[4] Apparently, then,
there is reflective consciousness with its twofold
aspect, because there are *objective* and *subjective* factors *before* consciousness arises. And this view of
Mr. Lewes' meaning is confirmed farther on. "And
what is this 'greeting of the spirit'? The metaphor
expresses that reaction of the sensitive Organism

---

[1] At p. 52, in § 11. Elements and Aspects of Phenomena.
[2] Problems, &c. Vol. II. p. 13.
[3] The same, p. 14.   [4] The same, p. 15.

upon stimulus, which is one necessary factor in every phenomenal result, since every phenomenon is at once object and subject."[1]

Then again, philosophers are said to puzzle themselves with the problem, "What is the connecting link between these opposites ? What is the bridge over which the object passes into subject, and cause into effect ? There is no bridge. The object is object-subject, the cause is the effect, the effect is the *causatum* (see Problem V. Chap. II.), the *natura naturans* is *natura naturata*, viewed under opposite aspects."[2] Viewed, Mr. Lewes says, under opposite aspects. But how comes it to be so viewed, and by whom? My answer is,—by *reflection*, which first perceives and then holds *together* the two aspects, not taking first one, then the other, like the two sides of the same ball.

But Mr. Lewes does not see this reflective action. He supposes us either to separate or else to unify. "The separating intellect detaches the Cosmos from the universal Existence, and then detaches Consciousness from the Cosmos, as it detaches a particular from an universal. The identifying intellect reverses this procedure, and sees in the primary fact of Feeling an implicit unity of the two Aspects which are explicit in Abstraction."[3] "The Twofold Aspect is therefore the alternation of abstractions."[4] *Alternation* observe. " But all these are abstractions which Reflection restores to their real unity."[4] And again, "The best modern metaphysicians, with rare exceptions, are now agreed that whatever may be the case with ultimate existences, the phenomena we deal

---

[1] Problems, &c. Vol. II. p. 38.   [3] The same, p. 19.
[2] The same, p. 18.   [4] The same, p. 22.

with are bipolar, on the one side objective and on the other subjective ; and these are the twofold aspects of reality."[1]

Notwithstanding, then, that Mr. Lewes has laid hold of a piece of metaphysical insight, he does not follow the method by which that insight was obtained. He is still scientist in the application of it, though the principle applied is metaphysical. Both things are requisite to a satisfactory treatment of philosophical problems ; that is to say, both the principle and the method. What Mr. Lewes does not see is this, that reflection is a principle which distinguishes, and therefore *enables* a separation of, the two aspects ; but that it is not reflection, but direct perception, which either separates into two objects, or unifies into a single object, the aspects so distinguished. Hence the difference between metaphysic based on reflection and science based on direct perception.

The case stands thus, as I shall very soon proceed to show. Reflection, arising in primary consciousness, is the first to perceive and distinguish, without separating, the subjective and objective aspects. Next, ordinary thinking creates, by uncritical processes founded on this distinction, a world which appears to consist of Persons and Things, that is, of separate *subjects* and *objects*. Thirdly, science adopts and perpetuates this conception, only making the reasoning about the separate objects more exact, by measuring and verifying. Finally, analytic philosophy, or metaphysic, returns to the inseparable character of the two distinct aspects. The discovery of Reflection as a process of consciousness, and the

[1] Problems, &c. Vol. II. p. 35.

analysis of that process, itself performed by an exercise of reflection, are the whole gist of modern philosophy. And philosophy for this reason comes later to its maturity than science, its object-matter being more recondite.

But what Mr. Lewes wants to do is this, to return in philosophy, not to the inseparable character of the two distinct aspects and to the method which so considers them, but to the undistinguished unity of *primary*, pre-reflective, consciousness; and to return to this as if it was an object of direct perception, a single existent, a conception of it which really presupposes reflection; he wants to have, in his conception of the universe, the unity of *primary* consciousness along with the definiteness of *direct*, two incompatible things, since primary consciousness is only *one* by virtue of being indeterminate. It is impossible for us actually to return to the state of primary consciousness; it is impossible for us logically to acquiesce in the separations of direct consciousness; the only course open to us in philosophy is therefore to accept the position of *reflecting* beings.

§ 2. I must now ask the reader's best attention to the analysis of the "moment" of self-consciousness or reflection, and the relation in which it stands to both the primary and the direct states of consciousness. This is the first task. Next comes its relation to the two fundamental modes of consciousness known as Perception and Conception, which are common to the three modes, primary, reflective, and direct. And thirdly, I shall attempt to show farther

Book I.
Ch. II.
—
§ 2.
Its relation to
primary and
direct modes.

the bearing of this double result upon the questions of the validity and the limitations of the philosophical method.

The first thing to be done is precisely to determine the thing to be analysed. I do so in this way. We all know what is meant by saying—I find myself having feelings and thoughts, and in the presence of objects around me. This total state of mind clearly contains the object to be analysed, the moment of self-consciousness itself. And first I observe, that there are in it three things, the person having the feelings and thoughts; the objects around him; and the feelings and thoughts themselves.

Now it is a well attested fact of observation, that infants have feelings and thoughts without having the perception of *themselves* as persons. There may be a series of feelings and thoughts in what is afterwards perceived as a person, which are not referred by the subject of them to himself as their possessor. A series of feelings and thoughts is therefore a condition of the perception of self, and can exist independently of that perception.

But it is not so evident that a series of feelings and thoughts can exist independently and as a condition of the perception of the other member of our analysis, the objects around us. Objects as such are often held to be *immediately* perceived, in the sense that no feeling and no thought can exist at all unreferred to an object, although at a later period the person, or subject, takes the place of the object for some feelings and thoughts, which are thence called *merely subjective*.

What proof can be given that this is the case, what proof is there that independent objects around

Book I.
Cn. II.

§ 2.
Its relation to
primary and
direct modes.

us are given to perception in and with any and every series of feelings and thoughts? It clearly cannot be regarded as a self-evident truth. For, first, the analogy of the case of the person is against it; and secondly, we can clearly represent to ourselves the possibility of the opposite alternative; we can clearly represent to ourselves a series of feelings and thoughts existing without their being referred to objects at the time, by the sentient being; although he may *after-wards* perceive that objects were their condition of existing.

The gradations in the scale of sentience, too, speak loudly for this opposite alternative. Low organisms may clearly have feelings of heat and cold, pressure, light, and so on, without referring these to independent objects around them. Organisms better endowed have more complicated series of feelings; comparison of feelings becomes possible; groups of feelings can be put together and distinguished from other groups. But this is a process not of feeling only but of *thought;* and still it has not been necessary to suppose any reference of these feelings, groups of feelings, or comparison of feelings, to independent objects.

If, then, we would avoid any unfounded assumption in our analysis, that is, any admission into our analysis as ultimate or unanalysable fact what is really analysable, we must begin by assuming no more than the series of feelings and thoughts *per se*, unreferred (by their subject) either to objects or to self. That is to say, we must begin with the third of the three things distinguished at the outset in the object-matter to be analysed. The series of feelings and thoughts *per se* is the groundwork of the whole;

Book 1.
Ch. II.

§ 2.
Its relation to
primary and
direct modes.

this is what I call *primary* consciousness; and this must be submitted to further analysis in order to see whether it will or will not furnish us with an account, or become the analysis, of the two other members of our object-matter, namely, objects and self. If it will, we may rest assured that we have reached the final and true account of those perceptions ; since any other explanation would involve some violent or unfounded hypothesis, as, for instance, of a special faculty of thought or of intuition, for giving us objects or persons *per se*, or of a special revelation of them to consciousness.

I say it is the groundwork of the whole, because it is clear that both objects and the self are closely mixed up with the feelings and thoughts of primary consciousness. Neither objects nor self ever appear *per se*, but always in connection with these feelings and thoughts. One is the object, the other the subject, *of those states of consciousness*. These are the known *data*, to which they are referred. In some way or other the perception of independent objects and the perception of a percipient subject supervene upon, or are developed out of, these primary states. There is no other alternative. For since neither objects nor self are given alone, *per se*, and also are not given along with the primary states originally,— there must be a time or "moment" of change, in which primary states of consciousness receive a modification and pass into states either of self-consciousness, or of consciousness of objects, or states which are both together: in which, therefore, these latter states are subsumed under primary consciousness, as a modification or case of it.

In what, then, does this modification consist?

Book I.
Ch. II.

§ 2.
Its relation to
primary and
direct modes.

What does the perception, in its lowest terms, consist of, which contains, in however incomplete and undeveloped a shape, the *differentia* of reflection? Or, to put the same question as one of history, so as better to keep before the mind the previous condition of primary states,—What is the first and simplest reflection which arises in the primary consciousness of an infant?

The moment of reflection being isolated and mentally passed in review by these questions, the answer, which is given by introspection, is simple. It is this: These thoughts and feelings are not only thoughts and feelings, but bundles of constantly connected thoughts and feelings, that is, "things." The connection between them belongs to them. Therefore they are *things*, as well as, and without ceasing to be, *states of consciousness*. They have a double aspect; that which was undistinguished has, I now see, a distinction into consciousness and object of consciousness.

If an infant could speak proleptically, could use a post-reflective and philosophical language, it is thus, I imagine, that he would express his first reflective perception. In short, *connected stability in feelings, which do not on that account cease to be feelings,*—this is the thought or perception which is the transition from primary to reflective consciousness. The percipient's own limbs and body are most probably the particular case in which this reflection arises, or about which it is made, in the history of the individual. But the perception is the same, whether we imagine it in its first historical instance, or take it from actually remembered experience of our own, in later instances, when it is mixed up with other experiences. Our consciousness of things is the perception

Book I.
Ch. II.

§ 2.
Its relation to
primary and
direct modes.

that *the feelings and thoughts composing them* are felt;
our consciousness of self is the perception that those
feelings and thoughts are *feelings and thoughts*. The
same perception, about the same states of consciousness, is at once a perception of the *existence* of things
and of the *existence* of feelings and thoughts. In
primary consciousness there were thoughts and feelings, but there was not the perception either that they
were things, or that they were thoughts and feelings.

But now observe this most important circumstance. Language like that which I have supposed
the child to use is not only not the language in which
the first reflection is expressed at the time, but it is
not the language held at all, even after reflection
has arisen, not the first actual post-reflective language. The first reflection is a perception already
left far behind, when the child comes to express himself in language. It is a forgotten condition of those
states of consciousness, of those perceptions, which
are the first to be expressed by him. When he learns
to say ‘I’ and ‘me,’ he has already learnt that percepts
are things; and his ‘me’ is a thing among things, a
percept among percepts. His earliest post-reflective
language is language expressing, not reflection, but
direct perception which has supervened upon it. If
he could speak analytically, his earliest post-reflective
language would be: This thing or group of percepts is ‘me;’ those other things or groups of percepts are ‘my’ percepts. For this reason, that his
own body has been by him classed with and separated
from other surrounding groups of percepts, that is,
feelings and thoughts, by primary consciousness, before the particular perception arises that one of these
groups, his body, *has* the feelings which constitute

that and the other groups of percepts. This percep-

Book 1.
Ch. II.
————
§ 2.
Its relation to
primary and
direct modes.

tion it is which is indicated by calling 'me' that group
to which the others belong, or round which they
seem to cluster, and which is always present when
any of the others are. The child's earliest post-reflective language is therefore that of direct or separative
perception from the very first; and it is only philosophy which traces, by analysis of consciousness, the
reflective perception which has preceded and made
possible his direct or separative perceptions, intervening between them and his primary consciousness.

We must be on our guard against the fallacy of
incidental circumstance, the κατὰ συμβεβηκὸς, in this
matter. When Ferrier says, "Along with whatever
any intelligence knows it must have some cognizance
of *itself;*" or when Jacobi says, in a passage quoted
by me in *Time and Space*, "I experience that I exist
and that something external to me exists in the same
indivisible instant;"[1] the cognisance or the experience of the "I" is *incidental* and posterior to the real
moment of reflection, which gives the distinction between *feeling* and *thing felt*. We experience that they
are feelings, but not that it is *I* who feel them. For
this there must be a supervening direct perception,
which unifies the feelings into an *I* or self, as the
Subject of them. The feelings and things felt are
perceived at once, by reflection in the same indivisible moment; and the feelings are what will afterwards be called *I*. But that the feelings, the *subjective aspect*, are a *Subject*, an *I*, or a *Self*,—this is not
perceived in that indivisible moment; but is the product of direct separative perception combined with

[1] *David Hume. Ein Gespräch. Werke,* Vol. II. p. 175. ed.
1812-1826.

BOOK I.
CH. II.

§ 2.
Its relation to
primary and
direct modes.

it. The true meaning of Subject and Object is, accordingly, subjective and objective aspect; that is what they are known as, in reflection alone.

Let us now endeavour to trace the method by which this direct and separative perception springs from the distinction of aspects drawn by reflection. Primary consciousness suffices to separate groups or bundles of percepts, existing simply as states of consciousness, from one another, and the body of the observer is one of these groups. It is that group round which the rest seem to cluster, which is present when any of the rest are, and which is also present when feelings are experienced which have no visible and tangible existence outside the body, or at any rate only an imagined one; I mean such as heat and cold, internal bodily sensations, appetites, desires, and emotions. Upon this state of perception reflection supervenes, whereby feelings and thoughts are distinguished as being at once feelings and thoughts as well as what we afterwards call "things." Two analyses have then to be combined, that given by primary consciousness into separate groups, and that given by reflection of every group into inseparable aspects. Some *hypothesis* has to be found which will render easy the holding together of these two analyses. This takes place inevitably, in obedience to the Law of Parcimony, which is the ultimate practical law, or motive, of all reasoning, the universal tendency to simplify different facts or theories by reducing them to a single common fact or theory. The hypothesis adopted is, that all feelings belong to the body, and that this "thing," which is already separate from other "things," is different in kind from them, inasmuch as it is the abode and source

Book I.
Ch. II.

§ 2.
Its relation to
primary and
direct modes.

of feelings; in other words, the body becomes a "person." The objective and subjective aspects are thus separated as well as distinguished; and this state of the matter is that which is expressed by the earliest post-reflective language. This separation of the aspects is complete, when even the 'me' is analysed, the body, which was part of it, again classed with "things," and an immaterial substance, the soul or mind, imagined as the subject of the feelings and the bond of their union.

We have, then, the *separation* between groups of percepts, given by primary consciousness; and we have the *distinction* between percepts and things, given by the first exercise of reflection. The attempt to combine these two kinds of perception, to put together the distinction between percepts and things and the separation between groups of percepts, results in the grouping of percepts together in the immaterial soul, and the grouping of things together in the external world; that is, results in direct or separative consciousness.

What has now been proposed as the analysis of primary, reflective, and direct consciousness, applies alike to the infancy of the individual and to the infancy of the race. The difference is, that the course of development taken by the race, that is, by individuals in the infancy of the race, is repeated by individuals in the maturity of the race under the guidance and teaching of seniors belonging to that maturity. But the separation of Persons and Things is the theory attained and held alike by all. It is the point of view from which the sciences start as their basis, and which they do not attempt to transcend. Direct and separative consciousness is the

Book I.
Ch. II.

§ 2.
Its relation to
primary and
direct modes.

presupposition common to all the methods they employ.

Psychology alone of the sciences is led to reverse some of the later steps of the hypothesis, the formation of which I have traced, those steps by which an immaterial substance is substituted for the body of the observer. In this reversal of the popular hypothesis, psychology of course is opposed to the upholders of the unreversed hypothesis, the theory of an immaterial substance. But it is still bound to the theory of some substance being requisite for the support of consciousness, though it finds this in the bodily organism, or rather in certain parts of it. It stands in the last resort upon the same basis as the immaterial substance theories stand, namely, upon the assumption of direct separative perception, and consequent separateness of consciousness and its objects. Both assume the existence of "things," and of "things" immediately perceivable by "persons" or conscious beings.

But here is the point in which psychology and philosophy differ; for in tracing consciousness back to its sources we come to states of it which are primary, undistinguished into objective and subjective; and the question is, how such states pass into those in which objective and subjective aspects are not only distinguished but separate. Psychology assumes a world of "things," and supposes these to impress another "thing," the sensitive organism, with primary feelings; but, since these feelings are not, at the time of being impressed, the subjective counterparts of the "things" which are held by science to have impressed them, some theory is requisite as to how they have become so; in other words,

Book I.
Ch. II.

§ 2.
Its relation to
primary and
direct modes.

how the world of "things," as we know it by science, has grown up out of the world of primary feelings as we knew it at the beginning of our knowledge.

What I maintain then is, that, unless we assume the "things" to be absolute existences, and the "persons" to be absolute existences perceiving them (which is the popular hypothesis), the passage from the primary feelings to the subjective counterpart of the objective "things" is only possible by supposing a reflective distinction of aspects supervening on the primary feelings, and preparatory to the separation of aspects in direct perception. In other words, the separation of aspects is impossible without a prior distinction of them as inseparables; and any other process lands us, sooner or later, in the hypothesis of absolute separate existences, which is the very same crude hypothesis from which both psychology and philosophy began by attempting to escape.

The order, also, in which primary, reflective, and direct, consciousness follow one another cannot be changed; the primary states must be the first, and reflective the second; and the direct cannot come before the reflective. For if the direct states are not the first,—and they cannot be so except on the supposition of absolute existences,—then neither can they spring immediately from the primary states, in which everything is both "feeling" and "thing" indistinguishably; since this would be to reverse the *datum* from which they are supposed immediately to spring, by denying "things" to be "feelings," which nevertheless they are in primary consciousness. There would be a direct contradiction between antecedent and consequent.

Nor is the experience of feelings which have no

Book I.
Ch. II.

§ 2.
Its relation to
primary and
direct modes.

visible and tangible existence outside the body, which we have met with above,[1] by itself sufficient to account for direct separative perception, as might possibly at first sight be supposed. It might possibly account for a separate subjectivity in the body, but not for a separate objectivity in the things outside the body. For these things were originally feelings as much as any other class, and the question is, How did they cease to appear so? This transition must have taken place *consciously*, there must have been a conscious renunciation of the belief in the subjectivity of "things," as being a mistake; a renunciation made in order to have an hypothesis to reconcile the two analyses; this hypothesis being that of Persons and Things. But to say that the transition took place consciously is to say, in other words, that an intermediate stage of reflection intervened, an intermediate perception of the two aspects as inseparable, although without the further knowledge that they were for ever inseparable, a knowledge reserved for philosophy.

When at length the whole process is completed, when direct consciousness has supervened upon reflective, then these two modes replace primary consciousness, which was the original of them both, take up into them and contain its states as part and parcel of themselves. There is primary, undistinguishing, consciousness no more; there exist in its stead the two modes, reflective and direct; reflective which envisages both aspects together, direct which envisages one aspect only, as separable from, and therefore also (since it is always in connection with it) as conditioning or conditioned by, the other

[1] At page 114.

Book I.
Ch. II.

§ 2.
Its relation to
primary and
direct modes.

aspect. And of the two aspects, thus considered as separable, the subjective is, or rather contains, the *conditiones cognoscendi* of all the parts of the objective; and the objective contains the *conditiones existendi* of all parts of the subjective. In science and philosophy we have no longer to do with the primary states of consciousness alone, which are a thing of the past, except as furnishing the content of the actual states, those of direct and reflective consciousness, which have sprung from and replaced them, and as supplying the proof of the relation between them. This is why, in Chapter I., I spoke only of the contrast between direct and reflective consciousness; because there I was engaged only upon the contrast between science and philosophy, of which this is the basis. And the same must, in philosophy, continue to be our main point of view; for, though the primary mode of consciousness still exists, it exists as continued into and combined with one or the other of the two modes direct and reflective. And it is its combination with direct separative perception that gives plausibility to absolutist tenets.

Mr. Spencer will furnish an illustration of this. In his Chapter on the Corollaries of the Universal Postulate,[1] having first called on his reader to "expel from his consciousness everything that can be expelled : so reducing his consciousness to its prespeculative state," he proceeds: "Now let him contemplate an object,—this book, for instance. Resolutely refraining from theorizing, let him say what he finds. He finds that he is conscious of the book as existing apart from himself. Does there enter into his consciousness any notion about sensa-

[1] Principles of Psychology, Vol. II. p. 437. 2nd edit.

Book I.
Ch. II.

§ 2.
Its relation to
primary and
direct modes.

tions? No: so far from such notion being contained in his consciousness, it has to be fetched from elsewhere, to the manifest disturbance of his consciousness. Does he perceive that the thing he is conscious of is an image of the book? Not at all: it is only by remembering his metaphysical readings that he can suppose such image to exist. So long as he refuses to translate the facts into any hypothesis, he feels simply conscious of the book, and not of an impression of the book—of an objective thing, and not of a subjective thing. He feels that the sole content of his consciousness is the book considered as an external reality. He feels that this recognition of the book as an external reality is a single indivisible act. Whether originally separable into premisses and inference or not (a question which he manifestly cannot here entertain), he feels that this act is undecomposable. And, lastly, he feels that, do what he will, he cannot reverse this act—he cannot conceive that where he sees and feels the book there is nothing. Hence, while he continues looking at the book, his belief in it as an external reality possesses the highest validity possible. It has the direct guarantee of the Universal Postulate; and it assumes the Universal Postulate *only once*."

There cannot be a better description of the state of consciousness which I have called direct separative consciousness than this, nor can there be a stronger proof how completely direct consciousness has taken the place of primary in scientist theories. I need hardly, perhaps, stop to point out that, in the third line of the above passage, a notion *about* the book, namely, that it exists apart from the observer, is mixed up by Mr. Spencer with the perception of the

Book I.
Ch. II.

§ 2.
Its relation to
primary and
direct modes.

book itself, as if it were a part of the observer's immediate consciousness of the book. But this notion
is as little a part of that immediate consciousness as
the notion that it is an image, or that it is connected
with sensations. After this, no one who knows anything of the besetting difficulties of these questions
will be surprised, that Mr. Spencer finds it necessary
to "transfigure" the "crude realism" which believes
it perceives a real book, into a realism "which simply
asserts objective existence as separate from, and independent of, subjective existence;"[1] into a realism
which "affirms neither that any one mode of this
objective existence is in reality that which it seems,
nor that the connexions among its modes are objectively what they seem;"[1] into a realism to which "the
Object is the unknown permanent *nexus* which is
never itself a phenomenon but is that which holds
phenomena together;"[2] in short into a realism to
which the Real is the Unknowable. The means, too,
by which this transfiguration is effected will be at
once obvious. A notion *about* the object of perception, the book for instance, is first mixed up with
that percept; and then this percept is dropped, while
the notion about it is retained. The notion "objective existence" slides into the percept, "book;" the
percept "book" slides out of the compound so effected;
and the notion "objective existence" remains alone.

It will be instructive to see, in another instance,
how an earlier writer of the same school of objective,
or non-Idealist, absolutism, an absolutist of pre-
Kantian type, arrives at his conclusion of a reality
behind phenomena; I mean the acute, suggestive,

[1] Principles of Psychology, Vol. II. Chap. XIX. p. 494. 2nd edit.
[2] The same, Vol. II. Chap. XVIII. p. 484. 2nd edit.

Book I.
Ch. II.

§ 2.
Its relation to
primary and
direct modes.

and often profound, Herbart. Open Herbart's *General Metaphysic*, at the first Chapter of the *Ontology*. He there shows, what is quite irrefragable, that the *data* of all knowledge are actual feelings, *Empfindungen,* and that the authority of these data cannot be either lessened or increased. The question is, *what* this authority imports? He then proceeds, " *The forms of experience* (be they contradictory or not) *cleave to the feeling.* But the feelings are not things but states. The matter of the data consists of feelings; *therefore no things are given, nothing real is given.* What do we know, then, of the real? Nothing. We will therefore make the assertion : Nothing exists. There is no existence."[1]

But this nihilism, he proceeds to show, is totally untenable. It can be asserted in words, but it cannot be retained in thought. We can find the way from the world to nothing, but we cannot find the *way back*, from nothing to the world. " In fact it is clear, that, if nothing exists, then also nothing must *appear.*"[2] There is, then, a real existence behind phenomena; " so much phenomenon,—so much indication of the corresponding real existence. (*Wieviel Schein, soviel Hindeutung aufs Sein.*)"[3]

Where lies the error, if any, in this reasoning? In this, that he *assumes* that there must be *things* as a reality. When he says: " The matter of the data consists of feelings; therefore no *things* are given, nothing *real* is given," the true conclusion (but for this assumption) would have been,—therefore *the feelings are the real.* Instead of this, as he *will have* a ground of phenomena, he draws, first, the conclusion

---

[1] Allgemeine Metaphysik, § 199. Werke, Vol. IV. p. 69.
[2] The same, p. 69.   [3] The same, p. 70.

Book I.
Ch. II.
——
§ 2.
Its relation to
primary and
direct modes,

of total nihilism, and then, from the absurdity of this, the conclusion of a real existence behind phenomena. His whole reasoning rests on the pure assumption, that there must be a *thing-like* ground of phenomena, something answering to the popular impression connected with the word *real*. He is mastered by a name.

Herbart's next proceeding is to distinguish, in the conception of "the existent which is the ground of the phenomenon," two things, our knowledge *that* it exists, and its own unknown quality. "The unknown is the quality; but our conception of the existent consists of something known and something unknown, its *existence* and its *quality*."[1] So far as the first of these two elements of our conception of the existent goes, Herbart is no doubt wholly in the right. He bases himself on Kant's true doctrine of existence not being a property or characteristic in existents, but something said about them, a positing, *Setzung*, as he calls it. And this notion, though not adequate, inasmuch as it gives room for the Idealist doctrine that the existence of things is due to the positing of them by the percipient, yet is not incompatible with the truth that existence is a reflective and not a direct perception. As Herbart puts it, "Here remains merely the conception of *that the positing of which is not reversed (aufgehoben)*. The recognition of the-not-to-be-reversed is the conception of existence."[2]

The error of Herbart's assumption shows itself in the second element of his conception of the existent, its unknown quality. There is really no need to think of any particular quality, in the things

---

[1] Allgemeine Metaphysik, § 200. p. 70.
[2] The same, § 201. p. 72.

BOOK I.
CH. II.

§ 2.
Its relation to
primary and
direct modes.

which are thought of as existing, attaching to them *as existing*, over and above their quality as phenomena. This they must have, for they must be feelings; when we are aware of *anything* as a feeling or complex of feelings, then we can think of it as existing with that feeling as its quality. But Herbart is led (as we have seen) to a quality belonging to existing things *as existents*, over and above the feelings which are phenomenal. This quality of existents as such, though unknown, is with him essential to their existence. He *repeats* the phenomena of the phenomenal world in a supposed real world which is its ground. " If we are *now* asked: *How are we to manage so as to posit anything as existent?* our answer is: Posit it in the same way as you are used to posit the things in the world of sense, when you are seeing them or handling them, or perceiving their sound or taste."[1]

There is this difference between Herbart and Mr. Spencer, which is practically immense. Herbart proceeds to construct his noumenal world of real existents, and to explain in detail the phenomenal world by means of them: but Mr. Spencer, satisfied with proving the existence of his, and its causal relation to the world of phenomena, devotes almost the whole of his philosophy to this latter world, without troubling himself farther with the noumenal, which he represents as unknowable; merely adding, that our consciousness of Nature under its unknowable aspect constitutes religion, as our consciousness of it under its knowable aspect constitutes science.[2]

But Herbart's whole philosophy consists in ex-

---

[1] Allgemeine Metaphysik, § 201. p. 71.

[2] For these expressions see his *First Principles*, p. 106. 3rd edit.

Book 1.
Ch. 11.

§ 2.
Its relation to
primary and
direct modes.

plaining the phenomenal by means of the noumenal world. As he says in a remarkable passage of his *Psychology as Science*,[1] "It is only necessary for this purpose" [of attaining real knowledge] "to find the conditions under which alone the phenomenal world can be phenomenal (*die Erscheinungswelt erscheinen kann*); in such wise that it would not be phenomenal, if these conditions did not exist." An important *only;* for whence but from the phenomenal world itself are we to draw our conceptions of what these conditions may be, nay even of what is meant by *conditions* at all? The *noumena* alleged by the theory, be they what they may, will be just as phenomenal as the phenomenal world which they are called in to explain, only without their concreteness; they will be abstractions of some sort or other, points without magnitude, lines without breadth, positions without space, forces without matter.

And even if the existence of such real noumena is admitted, they become useless for purposes of explanation in the very moment that they are seized by thought. Space and Time, as Herbart allows, are "the form of orderly holding-together in thought generally, be the object what it may."[2] Now, as he presently says, "every Real considered in itself alone is an *Absolute;* therefore, and for no other reason, the Real, in itself, is not in time, not in space." That is to say, in the very moment of seizing *the Real*, it vanishes from "the form of orderly holding-together in thought;" vanishes, therefore, from any theory or hypothesis into which we may have introduced it.

---

[1] Psychologie als Wissenschaft, § 31. Werke, Vol. V. p. 292. ed. 1850.

[2] The same, § 102. Werke, Vol. V. p. 507. ed. 1850.

Book I.
Ch. II.

§ 3.
Its relation to
Perception and
Conception.

§ 3. I now turn to the second question proposed above (p. 107), the relation of primary, reflective, and direct consciousness to the functions of perception and conception.[1] All consciousness, since it occupies time, is a continuous process, and one consisting of a series of states of consciousness, in which respect the discreteness of the process consists. Any single state in this series may be considered alone, and as so considered it is taken statically, as if at rest, although it is in movement since it occupies time. Any portion of the series, long or short, simple or complex, may be considered in this way, and so regarded it is a percept. But fix on any such portion by attention, for the purpose of connecting it with another portion, and it becomes a concept; an expectant percept or inchoate concept, until completed by the connection of the new portion. This completion effected, the whole is a concept in respect of its consisting of two portions connected together by *thought*; but is a complex percept, if considered as a single thing not yet brought into connection, by thought, with other percepts. It bears a double character, being a concept in respect of its containing a connection of *thought*, and a percept in respect of its non-connection by *thought* with other percepts.

The percept, then, is a more rudimentary state of consciousness than the concept; single percepts are the lowest empirical things in the analysis; a single

---

[1] It is not intended to imply that the term *function* is exclusively appropriate to perception and conception. All consciousness is a function of the organism; and so also, derivatively, are its modifications or determinations. Here the term is used to mark that there is an important difference between perception and conception, on the one hand, and primary, reflective, and direct consciousness on the other, which are properly called *modes*.

Book I.
Ch. II.
——
§ 3.
Its relation to
Perception and
Conception.

inchoate concept is a percept *plus* the purpose of connecting it with another percept, and a single complete concept is the connection between at least two percepts. Thus perception may be regarded as a primary function, of which conception is a derivative, and a derivative of a particular kind.

The priority of percepts to concepts is a prominent doctrine with Mr. Lewes, stated of course in his own way;[1] and the importance which he attributes to it, great as it is, cannot be considered excessive. The distinction originated with Aristotle, and is one of the corner-stones of philosophy. I shall devote a separate Chapter to its examination; but for the present purpose it is sufficient to say, that conception is that act in reasoning, (reasoning being a volitional action), wherein we pass on from one percept to another, thus either comparing, or separating, or combining them. Percepts are presupposed in conception, conception moulds or groups them, and the result is a more complex percept. Percepts are of all degrees of complexity, and the conceptions which are their connections equally so. A percept brought in this manner into connection with another percept is a concept; perception being the act of having percepts, conception the act of connecting them. Concepts are thus percepts in movement under the guidance of volition, and volition is the feature which distinguishes chains of conception from chains of percepts or images in spontaneous redintegration.

Now the stage upon which these operations take place, the mode of consciousness to which they belong, makes no difference in the functions of perceiving and conceiving relatively to each other. The same rela-

[1] See Problems, &c. Vol. I. pp. 101. 136. Vol. II. pp. 16. 69.

Book I.
Ch. II.

§ 3.
Its relation to
Perception and
Conception.

tion holds good between them, whether they occupy the stage of primary, of reflective, or of direct consciousness. And for this reason, that they arise from the deepest and most universal feature of consciousness, that of its occupying *time*. A percept is a portion of time filled with consciousness, a concept is this portion, grown, passed into the next portion. The movement of time, distinct with feelings, gives us necessarily both percepts and concepts; percepts so soon as we attend to what we feel, and concepts so soon as we attend to the connection of what we feel at one moment with what we feel at the next moment. All these conditions are found alike in primary, reflective, and direct consciousness.

When reflection first arises, it arises as a perception completing a conception which begins in primary consciousness. It is the perception of the double aspect in what was up to that time undistinguished. A movement begun in primary consciousness is completed in reflective, by a new character being observed in what we, after that observation, call its object. Primary consciousness ends by a conception, which is also in its completion the beginning of reflective consciousness, and the first perception in it. As a perception its content may be expressed by a proposition in which subject and predicate are co-extensive and convertible terms, a proposition in which the subject is reflected back again in the predicate, and the predicate in the subject. *Reflected*, observe, *figuratively* speaking. Not all convertible propositions are judgments of reflection in the sense of self-consciousness; but on the contrary, these judgments are cases of convertible propositions. The particular reflection that states of consciousness

Book I.
Ch. II.
———
§ 3.
Its relation to
Perception and
Conception.

are things, or that the Subject is its Objects, constitutes, as we have seen, the reflective mode of consciousness; and this reflection is expressible as a convertible proposition, or one the terms of which are coextensive. Thus convertible propositions are a wider class than propositions of reflection or self-consciousness. At the same time these are the final test of those, owing to their ultimately subjective character. Perception, then, is the rudimentary function in reflection as well as in primary consciousness; and reflective conception is a derivative from it. The same might easily be shown to hold good in direct consciousness also.

We have, then, the three modes of consciousness, primary, reflective, and direct. acting one and all in the two functions of perception and conception, and with the same content in all, though under different shapes. There is nothing in reflective consciousness which was not potentially in primary, and nothing in direct which was not potentially in reflective. And of these, the higher in potentiality and earlier in time governs the lower and later; governs it in the sense of being a prerequisite or condition which cannot be transcended; a limitation or condition *sine qua non*, of the lower and later. Reflection presupposes primary consciousness which it cannot transcend; and direct consciousness presupposes reflective with the same limitation. But the limitation is by no means of the same kind in every respect, in the two cases. Primary consciousness is a mere matter or ὕλη to reflective; but reflective is a law to direct, a law which we see cannot be overleapt or disobeyed, when once philosophical analysis has made us aware of its existence, a law which

Book I.
Ch. II.

§ 3.
Its relation to
Perception and
Conception.

we then see we have always been obeying though
unconsciously.

In the same way, perception gives the law to conception. There is nothing in conception which was
not potentially in perception ; a conception may itself
be the means of discovering new features, which will
be verified by, that is, actually found in, perception.
And this is the true shape of the Aristotelian dictum
repeated by Locke, *Nihil in intellectu quod non prius in
sensu.*

The primary function, then, perception, in the
reflective mode of consciousness, or briefly reflective
perception, is that mode or function of consciousness
which is the ultimate arbiter of all others, the bar at
which they all stand. In this mode, objects are envisaged as they are to the Subject alone, as they are
in their subjective aspect which is coextensive with
and inseparable from their objective ; and at the same
time as they are in themselves, in their nature or
analysis, and not in their connection with other objects in consciousness, which is their genesis or their
history. As percepts they are looked at straight and
for themselves ; as percepts of reflection they are
looked at in their subjective aspect, or as they are
immediately known to the observer ; and thus the
nature of the thing perceived, as it is *per se*, is, in
reflective perception, identically the same as that
nature which is immediately known to the observer.

We have long had the distinction between percepts and concepts, and also the corresponding one
between the acts of perceiving and conceiving ; and
philosophy has long been familiar with them, under
these or other names. But it may be asked, have
we any names to apply to percepts and concepts as

Book I.
Ch. II.

§ 3.
Its relation to
Perception and
Conception.

themselves objects of reflection, or, what is the same thing, to their subjective aspect in contradistinction to their objective? If reflective consciousness has percepts and concepts, the mind must have attitudes towards them as objects, some nameable attitudes corresponding to them. We are continually reflecting on our percepts and our concepts, and making them objects although states of consciousness. In making them objects the mind supplies them with a subjective aspect, yet in so doing always keeps to the lines laid down by its original character of being feeling in time, the lines of perceiving and conceiving. Is there, then, any name expressing this reflective action of the mind, the subjective side of the result expressed by "percept" and "concept"? Percepts and concepts are not found ready made in nature, they are consequences of certain mental operations which nature's laws determine. Yet particular percepts and concepts come partly from volition; interest attracts our attention to some things instead of others; and the percepts we begin with in any train of reasoning are thus determined. It would seem that this subjective side of the result, that in which volition has some share, requires naming as well as the objective side. The *process* common to both is named already, perception and conception. But the subjective side of the result, the mental attitude to percepts and concepts, has this no name? It ought to have, if what was said above of the continual regress into subjectivity, the continual development of reflective consciousness, is true. And what is more important than the naming them, the distinguishing them consciously as the subjective aspects of percept and concept, is essential to their adequate employ-

Book 1.
Ch. II.

§ 3.
Its relation to
Perception and
Conception.

ment. They ought to be brought forward, and made to play a part as fundamental in modern philosophy, as their objective aspects, percept and concept themselves, have played, and would continue to play alone, were they not, in the regress of reflection, supplemented by their subjective aspects.

It is to these subjective aspects of percepts and concepts that the old names, First Intentions and Second Intentions, in the new meaning which I gave them in an earlier work,[1] properly apply. These names express attitudes of the mind, of which percepts and concepts are respectively the objects. I was not sufficiently alive to this, when I made First and Second Intentions one of my bases in *Time and Space.* I connected them indeed with percepts and concepts, and showed the compatibility of the two pairs, as will be seen more particularly from the second passage cited; but the notion of the one pair being the opposite aspect of the other pair, and necessarily arising as such in reflection, did not occur to me. Nevertheless it flows as a legitimate consequence from what I there said; and in this way throws back light on the method of that former work, and renders it an indirect support. It shows that I was really working on the basis of reflective consciousness; and my procedure there may serve as an instance of what I now call philosophical method as distinguished from scientific.

---

[1] Time and Space, § 10. and § 46.

BOOK I.
CH. II.

§ 4.
Its importance
as sole source
of verification

§ 4. Having thus analysed reflective consciousness and shown its connection with primary and direct, as well as with the functions of perceiving and conceiving which are common to all, I now come to the final part of my task, and shall endeavour to show the bearing of this double result on the validity and the limitations of the philosophical method. This bearing depends on the character of reflective perception; as perception (not conception), it is that mode of consciousness in which all final verification of facts as well as theories is performed; as reflection, as recalling and repeating the phenomena of primary and direct consciousness, it is restricted to those phenomena for its contents, and is not a mode or an organ of new or transcendent discoveries. It is a reexamination of the phenomena of primary and direct consciousness, under the guidance of the principle of examining their objective and subjective aspects in conjunction with each other, which is a method only possible in reflection. Reflection first makes the discovery of the double aspect and then applies it, and the continued and methodical application of it is metaphysic.

All final verification of facts is thus performed by reflection. Scientists on the contrary maintain, that all final verification of facts is performed by direct observation. The reason of this contradiction is, that the finality point is differently placed by the two disputants. Scientists stop short at a final point which is not really final; and by really final I mean not an absolute end, but the furthest point attainable by us. But even here there is a difference between scientists themselves. One class, the stricter or more positivist among them, would urge that, having once

Book I.
Ch. II.

§ 4.
Its importance
as sole source
of verification.

arrived at an agreement as to the denotation of the terms, say, for instance, matter, motion, and force, we can test any propositions involving these terms by means of observation and experiment; the facts directly observed either will or else will not answer to our predictions; and thus direct observation verifies the truth or falsity of the propositions. Their finality point is placed at the agreement already arrived at about the things as *denoted* by the terms which they use; and they refuse to go any farther; it is enough for all purposes of science, for all purposes of practice, to start from an agreement as to the denotation; the facts, they say, are what concern us; why enter upon a futile question as to the *nature* of the things which the terms denote?

Another school, that for which Mr. Lewes has declared, urges on the other hand, that the *connotation* of such terms as these must be examined, the *nature* of the things denoted by them. Only thus, they maintain, can the belief in fictitious entities underlying them be destroyed; a belief which, as Mr. Lewes well urges, is not confined to metaphysicians, but is equally rife among physicists and physiologists, being founded on the tendency to suppose a separate entity wherever there is a separate name.[1] He argues therefore, irresistibly as I think, that it is requisite to examine and analyse what we really mean when we use such terms as matter, motion, force, and so on. Nor does he merely argue that it ought to be done, but he has done it. His two "Problems," IV. Matter and Force, and V. Force and Cause, in Vol. II., leave little, if anything, to be

---

[1] Problems, &c. Vol. II. p. 223.

Book I.
Ch. II.

§ 4.
Its importance
as sole source
of verification.

desired in exhaustive demolition of entities lurking under these names.

Wherein, then, it may be asked, is there cause for dissatisfaction here? Is not everything performed here, by way of analysis, that can be performed? How can Mr. Lewes be said to place his finality point too near? In this, I reply, that he does not recognise the nature of the method which he himself employs in this analysis, consequently cannot generalise it, consequently does not see that it is not a scientific but a metaphysical method. In examining into *what we mean* by such terms as matter, motion, force, he is really exercising reflective consciousness. He is comparing representations, remembered states of presentative consciousness, either with actual presentations, or with representations of them as such. In other words, the moment we begin to enquire into the *meaning* of an agreed on denotation, we begin to reflect; we begin to repeat the process by which that agreed denotation was itself arrived at; we go back into the analysis of the denoted thing, not as it will act and be acted on by other similarly denoted things, but as it is perceived and conceived by us; we are analysing it into its *subjective* elements.

The reason and necessity of this process are evident; it depends on this, that names are agreed on to *denote* objects of common enquiry, before the reasons, for which each party to that agreement uses the names, are agreed on; in other words, the connotation of the names is still unsettled, different persons having possibly different connotations in view for the same name. And in this way the agreement on the connotation of names involves a more searching analysis of the nature of things. This question cannot

Book I.
Ch. II.
§ 4.
Its importance
as sole source
of verification.

be settled at once, by asking what the name *originally* connoted ; for this is the main question itself in another form, and is the very thing *not* originally agreed on.

I will take an instance from Mr. Lewes' argument against Primary and Secondary Qualities of Matter :[1] "Having these points clearly fixed, let us begin our investigation of the Properties of Matter by interrogating Experience, and enumerating what are the special feelings grouped in the abstract symbol. The positive qualities are, of course, all those qualities which we perceive in substances. To Perception, Matter is those qualities, and it is nothing else. We need not here consider the argument which proclaims that Matter is something underlying and not identical with these qualities; our present purpose is with the qualities themselves. The object now held in my hand, seen and felt by me as coloured, figured, resistant, rough, smelt as fragrant, tasted as acid-sweet, I and my fellow-countrymen call a strawberry, and all men consider to be a substance, or portion of Matter. Reflecting on my experiences of other substances, and comparing these with the strawberry, I notice that it differs from them and agrees with them in the kind of feelings excited, and in the degrees of excitation. I class these feelings, and call the one set particular qualities, the other set general qualities; and on further comparison I find that, of the general qualities, some are universal."

Now without in the least insisting on the use of the word *reflecting*, in the above passage, which I cannot suppose is intended to be used in my technical sense of it, I think it is clear that the interrogation

[1] Problems, &c. Vol. II. p. 272.

of experience described involves what I mean by reflection. For it includes 1st, the identification of feelings with qualities, and 2nd, the representation in memory of former presentations, and subsumtion of them under that identification. The reflection consists in the identification of qualities with feelings; the generalisation extends the range of one such case of identification to all. And the result is, that all qualities of matter are feelings or relations of feelings. The word *strawberry* is the denotation, of which the connotation is sought; but sought, not in terms of its connections with other objects, but in terms of its meaning for consciousness.

Book 1.
Ch. II.
——
§ 4.
Its importance
as sole source
of verification.

This same method which Mr. Lewes employs, without recognising its true nature, is applicable, when fairly recognised, to other cases which he has left among the puzzles which are the legacy of science to philosophy. And this is why I have thought it necessary to insist so much on the present point. Mr. Lewes has overstepped the magic line of demarcation between science and philosophy, without knowing that he has done so; by the aid of an unacknowledged metaphysical principle he has succeeded in solving some of the problems of philosophy, and there is no course left for him, if he would solve the remainder, but to recognise and apply that same principle again; and I hope to show in future Chapters that it is fully adequate to their solution. But then what will become of his pretension to solve such problems without recourse to metaphysical and solely by scientific method?

We can now see the truth with regard to the contradiction with which we began: "All final verification of facts is performed by reflection;" and "All

Book I.
Ch. II.

§ 4.
Its importance
as sole source
of verification.

final verification of facts is performed by direct observation." On the one hand reflection is clearly included in it; on the other direct observation. But
reflection is the *method*, and includes direct observation as the *means*. This is the solution of the contradiction. Reflection is in science, not exclusive of but
superinduced upon direct observation; it is a riper
mode of consciousness in which direct observation
becomes a means and an instrument; the whole, of
which the less ripe mode is a part. The importance
of the instrument to science, together with the ontological dreams fathered upon reflection by too many
philosophers, have blinded both philosophers and
scientists to the true state of facts.

The errors of those who adopted the principle of
reflection, the "too much" of ontologising philosophers, are no argument against the principle itself.
The principle becomes gradually clearer and more
definite, sifted gradually from its accretions, in and
by its continual application to questions of speculation. Two periods in the history of its development
in individuals may be broadly distinguished, the first
beginning with the dawn of self-consciousness, the
second beginning, where it arises at all, with the
recognition and application of the fact of self-consciousness to the facts and problems of existence. In
both periods there is a growth and development of the
principle. We are not philosophers when we first
become reflective; we are not full-grown philosophers
when we first begin to apply reflection to philosophical questions.

The same two periods may be distinguished in the
history of philosophy itself. The principle appears
very early in Greek speculation, with Parmenides.

Book I.
Ch. II.

§ 4.
Its importance
as sole source
of verification.

But such statements of it as he gives, τὸ γὰρ αὐτὸ νοεῖν ἐστίν τε καὶ εἶναι, for instance, are to be paralleled with the dawn of reflection in individuals, for individuals are not philosophers merely in virtue of their self-consciousness. The principle must be recognised, and not only recognised but applied after recognition. Then only do individuals, then only does the race, begin to philosophise. Aristotle clearly recognised the fact of self-consciousness; for instance, in the words: καὶ αὐτὸς δὲ αὐτὸν τότε δύναται νοεῖν.[1] He recognised also the continued exercise of it, νόησις νοήσεως, as the highest intellectual energy; but he did not recognise, still less apply it, as the evidence and the only evidence of existence.

Nor must we suppose that this fact was less familiar to the mediæval schoolmen. We have, for instance, a remarkable passage of Scotus Erigena; put however into the mouth of the *Discipulus*, not the *Magister*, of the dialogue: "Scio enim me esse, nec tamen me præcedit scientia mei, quia non aliud sum, et aliud scientia, qua me scio; et si nescirem me esse, non nescirem ignorare me esse: ac per hoc, sive scivero, sive nescivero me esse, scientia non carebo; mihi enim remanebit scire ignorantiam meam. Et si omne, quod potest scire se ipsum nescire, non potest ignorare se ipsum esse; nam si penitus non esset, non sciret seipsum nescire: conficitur, omnino esse omne, quod scit se esse, vel scit se nescire se esse."[2]

Although it is clear that Erigena is here most

---

[1] De Anima, III. cap. 4.

[2] De Divisione Naturæ, Lib. IV. Sect. 9. Fol. 776. B., ed. II. J. Floss, 1853. In the Migne series. And see another passage to the same effect, and in the mouth of the Magister, Lib. I. Sect. 48. Fol. 490. B., cited by M. Hauréau, *Hist. de la Phil. Scol.* p. 183. ed. 1872.

Book I.
Ch. II.

§ 1.
Its importance
as sole source
of verification.

forcibly and explicitly enunciating the Cartesian *Cogito ergo sum*, before Descartes; it must yet, I think, be admitted on the other hand, that it was not till Descartes erected the clearness and distinctness of the evidence with which he perceived that his own consciousness involved his own existence into a norm of certitude,[1] that consciousness as such was distinguished from all its objects, and that thereby philosophy was distinguished from science; and it was not till he proceeded, as he immediately did, to enquire into the conditions of that evidence, and to separate[2] the immaterial soul, as the subject of consciousness, from the material or extended world, of which the living body was a part, that the basis of a physiological and scientific psychology was laid.

Psychology was not the discoverer of the principle of reflection, but a creature of it; it is not a principle of psychology but of philosophy. Nor indeed has psychology hitherto discovered either what physiological process is the condition, or what special portion, if any, of the nervous organism is the seat, of reflection; the discovery is reserved for the future. The philosophical character of the principle is also shown by its *range*, even as employed by those who do not recognise its nature; since it is employed, as we have seen it is by Mr. Lewes, to prove propositions about matter and force, that is, about the objects of the external world at large, the objects of all the physical sciences. The principle of reflection is a golden thread running through the whole of

---

[1] De Methodo, Sect. IV. "*Post hac inquisivi —— distincte percipimus.*"

[2] Principia Philosophiæ, Pars I. Sect. 11. 12. Pars II. Sect. 1. Pars IV. Sect. 189. And the Passiones Animæ, Pars I. Art. 2. 3.

Book I.
Ch. II.

§ 4.
Its importance
as sole source
of verification.

thought, and is thus the peculiar basis and instrument of philosophy, as direct consciousness is of the special sciences. Just because it is employed in every science, being the basis of one of their processes, and that the most essential, the process of verification; just because it aids to perform this essential function in them; we are entitled to argue the validity of the method which it furnishes to philosophy, which is nothing but the one general enquiry which is based on and completes the results of the special sciences taken collectively.

Let us now glance at an objection which may be raised to the validity of the philosophical method, the method of reflection or subjective analysis. There is nothing which thoroughgoing scientists so much dislike as the word *subjective*. It is unfortunately the only thing which philosophers cannot gratify them by renouncing. Scientists take hold of the admitted truth, that immediate certainty is only for the individual mind, which is immediately certain of its own feelings, but has no such immediate certainty of the feelings of other minds. And they argue that, so long as we keep to subjective analysis, we are confined within the limits of the individual's feelings, and cannot reach any knowledge which is valid and known to be valid for other individuals, or which, in other words, has *objective* validity. It is only, they say, those propositions which can be communicated and made intelligible to others that are the vehicles of a common knowledge, valid for all men, and therefore properly called objective. In other words, they argue that pure subjectivity is pure individualism, and therefore is possibly wholly chimerical, and at any rate cannot be brought to a test of whether it is chimerical or not.

Book I.
Ch. II.

§ 4.
Its importance
as sole source
of verification

My answer to this reasoning is, that it either proves too much, or else proves nothing, according to the meaning attached to the term *validity*. It proves too much, for the objector's purpose, if we take validity to include subjective and immediate certainty, for then the special and positive sciences will not have validity, unless a basis of subjective and immediate certainty is made out for them by the philosophical method. In this case, the sciences require something which only philosophy supplies, namely, the full proof of the validity of their knowledge. On the other hand, it proves nothing, if we take validity to consist in the knowledge being communicable to, and tenable by, all men, for this is not the exclusive characteristic of the special sciences, but is possessed by the truths of philosophy also. No difference is then made out between philosophy and science. In adopting the latter definition of validity, scientists adopt too low a standard of it, inasmuch as they renounce the demand for immediate certainty. Philosophy, on the contrary, shows how a basis of immediate certainty is obtained for the communicable objective knowledge of both itself and science. It aims at securing validity of a higher order than science contemplates as being even possible.

In proceeding to develop this answer more at large, I remark in the first place, that the scientist reasoning unwarrantably *separates* the two aspects, objective and subjective, not satisfied with distinguishing them: and it is only by this separation that subjectivity can be identified with individualism. Admitted that immediate certainty is only for the individual mind, still there cannot be an immediate

certainty in the individual mind which is not an
equating or an identifying of an objective with its
subjective aspect. This identifying of opposite or
distinguished aspects is what immediate certainty
consists in. This is verification, and this is performed only in individual minds.

But now comes the question, How can the individual be sure that the objective aspect, which he
has thus identified with its subjective aspect in his
own mind, is identified or identifiable with an objective aspect, similarly identified with its subjective
aspect, in another mind; how can he know that an
object for him is an object for others, is a common
object, or, as the scientists put it, is really *objective*.
This question science passes over, content to begin at
a lower point, namely, with facts which have already
got, somehow or other, that is, which are *assumed to
have*, this kind of objectivity; and just this question
it is which philosophy attempts to solve.

Now, however the knowledge of this kind of objective validity may be attained (a point which I shall
come to shortly), at any rate it cannot be given *immediately* by subjective observation alone. Here is
the place to apply the admitted truth, that subjective observation is confined to the individual himself.
Whatever claims may have been made on this point,
as, for instance, that we know immediately certain
particular truths to be valid for all intelligences, or
that we have an immediate intuition that some truths
not only are but must be true eternally,—all such
claims must be given up. It is not on this point, on
the question how we know *our* objects to be objects
common to *other* minds, that subjective observation
can be appealed to. For clearly subjective observa-

Book I.
Ch. II.

§ 4.
Its importance
as sole source
of verification.

tion, where it has immediate certainty, is reflective; and to attribute this knowledge to it is to suppose it not reflective but direct: it is to assert *reflective* knowledge between *separate* minds, a contradiction which it is the business of philosophy to expose, no matter whether in the mouth of scientist or philosopher. Scientists, then, may dispense with attacking this position of their brother scientists, which has been surrendered to them, unless they should be writing *ad populum*.

The immediate certainty of objective knowledge, where objective means common to all individuals, is *not* what metaphysicians mean by the phrase; for *objective* means, with them, objective aspect of an individual's subjective knowledge; neither is it what is intended when reflection is said to be the ultimate process of verification.

But now for the question *how* the transition is effected from the individual's knowledge to a knowledge common to him with other individuals, *how* he knows that his object is their object. In the first place I remark, that this question does not concern the philosopher more directly or closely than it concerns the scientist. Indeed in one respect it seems to concern the scientist more closely, because he maintains that this community of the objects is the distinguishing mark of science, (we have seen that it begins with the *assumption* of it), and gives it a claim to validity which is wanting to philosophy. But the truth is that it concerns both alike, inasmuch as it is a necessary feature in all knowledge, so far as that knowledge is systematic and available for use. I remark farther, that it is the proper function of philosophy to point out the

Book I.
Ch. II.

§ 4.
Its importance
as sole source
of verification.

answer to this question, inasmuch as it is a question approachable only by methods of subjective analysis, a question which concerns primarily the "how" of *knowledge*, and not the "how" of *things*. And thirdly I remark, that it would be of no practical consequence to the special sciences, though this question were not answered at all ; they would be equally useful and equally applicable. In these two respects, then, the question concerns philosophy more than it concerns science.

How do we know, in what way do we reach a legitimate certainty, that our objective aspects are the same as those of other individuals? The answer is extremely simple. It is by inference from what we perceive other individuals do or say or look like, in presence of ourselves and our objects at once. We perceive them doing and saying just what we do and say in the same circumstances ; in short, *re-acting* just as we perceive ourselves doing ; our own re-action follows from our own feelings ; we infer similar feelings in them, and similar feelings corresponding to the same objects. Such, in general terms, is the process ; so simple and so obvious, that many will be surprised to find it claimed as philosophical, and special importance attached to it. But the question is, what is its analysis? It is inference, resting on subjective observation, not knowledge immediately given by subjective observation. Subjective observation, tested by reflection, is here employed to interpret a class of objects, namely, other individuals, their sayings and doings, and to infer of another class of objects that they are common objects to us and to the other individuals.

When, therefore, a scientist urges, that science

Book I.
Ch. II.

§ 4.
Its importance
as sole source
of verification.

begins with facts agreed upon, predicts facts which
have a definition, and verifies the prediction by the
occurrence or non-occurrence of the fact as defined;
and contrasts this capability of verification with the
incapability of verification which, as he maintains,
distinguishes subjective observation; the reply must
be made, that this agreement as to facts and their
definition presupposes a long continued process of
attaining the agreement, in which process subjective observation has been employed at every step.

Nor is there any difference in this respect between physical objects and mental ones, such as
images, thoughts, and emotions. The only difference is one of degree in remoteness or in complexity;
physical objects have been the first to be described
and defined, because they are objects of so many
different and for the most part external senses;
mental objects have been later and less perfectly
described and defined, because they cannot be seen
by the bodily eye, or touched by the bodily finger,
while present to the consciousness of one individual
or of several. But the same process of attaining
agreement respecting them, of becoming certain that
our objects are the same as those of other persons, is
employed in their case also.

The method in the case of mental objects is
this;—we point the thing out by means of its connections or its accompaniments: grief, for instance,
we point out as the mental feeling which arises in
such and such circumstances, or as similar to such
feelings, dissimilar from other such; just as we point
out a physical object by its connections, as 'this
apple which I hold in my hand.' Names are given
and agreed upon, in the two cases, in manners pre-

Book I.
Ch. II.
——
§ 4.
Its importance
as sole source
of verification.

cisely similar. We build the names for mental objects out of the names for physical, just as we point out the mental objects by means of the physical circumstances attending them. In no case do we come to a name with an *agreed* connotation as ultimate; always the ultimate agreed name is denotative or designative, that is, it *is used* as a denotation, having indeed another meaning which is not named, its meaning to the individuals separately; its connotation to them, which constitutes it a *general* term. The name acquires its agreed meaning by convention, being originally a sound attached in common by several persons to feelings (which are its connotation to each of them individually) arising from the same object. The true or agreed connotation of the name is then arrived at by analysis of the object which, when first attached, it was employed to denote.

The two processes, then, analysis of physical and analysis of mental objects, cannot be sundered. Subjective observation underlies them both, and reflection verifies them. The attainment of agreement in defining objects common to all observers is the prerequisite of science; wherever that is done, whether in physical objects or in mental, as in the case of the moral sciences, there and at that point science begins. But the method by which that agreement was attained is not finished and done with, because a science or two has been founded by means of it. The whole of possible knowledge must (*per impossibile*) have been reduced to science, before that could be the case; and even then the *logical* primacy of reflection would continue, though its *productive* energy would have ceased.

Observe, then, the fundamental difference of

Book I.
Ch. II.

§ 4.
Its importance
as sole source
of verification.

method necessitated by this difference of the two
starting points of science and philosophy. Science
starts with an agreement already attained as to what
object shall be called by what name, that name being
a denotation; and its aim is to find what relations
obtain between this object and others similarly
agreed upon and denoted; and relations, too, are
objects agreed upon and denoted in the same way.
Its process is, in this respect, synthetic or constructive, and its ultimate units are the denoted objects
and their denoted relations. Philosophy on the other
hand starts with a subjective observation of objects,
whether these have been already agreed upon and
denoted, or before they have been so; not however
with the purpose of changing denotations which have
been agreed upon, but with that of fixing or enlarg-
ing their connotation; and if no denotation has been
established, then of choosing one with a suitable connotation attaching to it. Its method therefore is
analytic; and analytic without assignable limit.
The terms 'self,' 'ego,' 'I,' are themselves denotative
terms, and as such submit to the same process of
analysis providing them with a connotation, as the
units of physical and of psychological science submit
to. This subjective and analytic procedure of philosophy is what causes a phenomenon which has been
often noticed, particularly by J. S. Mill in a famous
passage of his *Logic*, that the great disputes in philosophy have so often taken the form of disputes
about the definitions of terms, the definition of Justice for instance. The very nature of philosophy is
to enquire constantly about the *meaning* of terms,
and to do so by analysis of the objects which we
represent to ourselves when we use them.

BOOK I.
CH. II.

§ 4.
Its importance
as sole source
of verification.

This is why I said at the beginning of Chapter I., that distinctions not definitions must always be the true starting point in philosophical method. Substantially the same thing has been clearly seen and forcibly expressed by Herbart, and that too in the very first sentence of his *General Metaphysic*,[1] so forcibly as perhaps even to overstate the case against himself as metaphysician. " Metaphysic," he says, " has *no foundation truths;* it has *foundation errors* instead." Mathematic has its axioms and definitions as basis; but metaphysic begins the world penniless. It has to discover by careful observation of the chaos, what, if any, is the real law of thought and clue to truth. Facts subjectively observed are all it has to trust to.

On the whole, then, I conclude, that we have in philosophy a valid method, based upon a valid principle, that of reflective consciousness. ' A thing *is* what it is *known as*,' this is the principle in its objective formula; ' the objective and subjective aspects are inseparable,' this is its subjective formula, expressing the way in which the same truth appears in reflection itself. This is no truth of science. When have we not heard scientists reply to this or similar statements, ' But you *don't* know the things;' or ' The question is what things *are* known as'? When have we not heard scientists tax the method of subjective observation with being chimerical and incapable of verification; and scornfully advise a *preliminary* agreement on the meaning of terms employed by it? Scientists have neglected the principle of reflection, from its being too obvious, too familiar. They have " idolised" verification, without seeing that the method of subjective observation

[1] Allg. Metaphysik, Einleitung: Werke, Vol. III. p. 66.

Book I.
Ch. II.

§ 4.
Its importance
as sole source
of verification.

did not *require* verification only because it *was* verification, was itself the operation of verifying, was the performance of that which they demanded that it should appeal to as a test. They idolised verification without having analysed it. They left their processes in the concrete as well as their terms, their methods as well as their objects.

And now comes Mr. Lewes and, employing this method without acknowledgement, claims to have solved the problems of philosophy by the methods *of science*. Science, which has so long disdained the labours of philosophy, at last silently appropriates them. Is this an acknowledged method of Positivism? I know not how the proceeding can be, I do not say justified, but explained, except by the influence of a certain fallacy which invests itself in countless forms, the fallacy of attributing all virtues to what we hold dear, all defects to what we dislike. Scientists hold science so dear that they attribute all truth to science, all error to what is not science. Whatever is true, whatever shall be hereafter proved true, must they think belong to it, its claims are imprescriptible. True enough, *if* scientific philosophy is included in science; but not true of science as separate from philosophy. As between philosophy and science, some truths belong not to science but to philosophy. Of this number is the principle in question.

So far, then, as to the validity of the method of reflection as a method of philosophy. I now turn to its limitations. These are drawn from the phenomena of primary consciousness, appearing as they do in the objects of direct, which it is the office of reflective to criticise. Reflection is no "faculty," independent, with laws and methods of its own, bringing into our

Book 1.
Ch. II.

§ 4.
Its importance
as sole source
of verification.

horizon as it were a new kind of feelings. It is a remoulding of the feelings already given in primary consciousness. This very relation of it to primary, which furnishes all the matter of direct, consciousness, which gives it the power of control and makes it the medium of verification, also fixes its limitations and defines its sphere.

First as to its origin out of primary consciousness. It arises in redintegration or representation of feelings. Feelings again are different modes of sensibility; there are feelings of external sensibility and of internal, and there are also sentiments, emotions, and passions, all of which are sensibility of a more complex kind; and there is a class of feelings which include a sense of effort or tension, among which desire, volition, and choice, are to be named, along with the more complex feelings into which these enter as constituents.

It is most important to remark that reflection is not *inner sense*, or inner feeling, as some of these modes of sensibility have been named. This way of looking at the matter would simplify it most conveniently, if only it were in accordance with the facts which I have insisted on in this Chapter. What can be more attractive than the following apparently plain statement of Czolbe :[1] " The single domain of our consciousness divides itself into two halves, the perception on one side of the external world, on the other of our own person. If our attention is fixed on the first half, we have world-consciousness, if on the second self-consciousness." Self-consciousness is then explained as being no special mode of consciousness, but consciousness with a particular object, namely,

[1] Grundzüge einer Extensionalen Erkenntnisstheorie, p. 40.

Book 1.
Ch. II.
——
§ 4.
Its importance
as sole source
of verification.
inner experiences in connection with the external perception of our bodily figure.

This is an instance of the *direct* mode of philosophising, against which I have contended. Were this the truth, reflection would be a mode or among the modes of primary consciousness. It would then be, not a mode of control or verification, but an original source of facts, that is, of feelings and phenomena. That which is now our safeguard would then become something to be guarded against. Inner sensibility in all its modes must be submitted to reflection, and each class of them watched in its different phases and its different connections.

Next comes the question as to the sphere of reflection and the limitations of its control. Mr. Lewes' Chapter on *Intuition and Demonstration*,[1] in which he argues against the so called Intellectual Intuition, furnishes an excellent instance of this control being exercised by reflection. Having enumerated examples of sensible intuitions, space, time, motion, quantity, &c.; of rational intuitions, substance, cause, equality, &c.; and of moral intuitions, freedom, responsibility, duty, &c.; he proceeds: "The validity of all these intuitions depends on their reduction to identical propositions; in other words, whether the relations *are* what we *see* them to be. The possibility of error lies in the possibility of our supposing that we see what we only infer." And again of the demonstrations by which intuitions are brought to the test: "Demonstration is not an instrument of discovery, but a means of control. Intuition is seeing; Demonstration is showing. *What* is seen and *what* is shown, may be illusory; they are

[1] Problems, &c. Vol. I. p. 368. et seqq.

BOOK I.
CH. II.

§ 4.
Its importance
as sole source
of verification.

only proved to be objectively valid when each infer-ence has been reduced to its corresponding sensible."[1]

All this is excellent and admirable. I do not remember to have seen anywhere the essential fea-ture, in which the verification of intuitions consists, more closely and clearly delineated, the purging of the imagination from false inferences, wrongly sup-posed to be immediate objects of perception. All I would remark is, what Mr. Lewes omits, that the demonstration here described is a process of reflective consciousness, not of direct merely; that the "iden-tical propositions" he speaks of are identical proposi-tions of a particular kind, namely, of reflection, and that this is essential to their verifying, or *ultimately* demonstrative, power.

I will now anticipate a reply which will probably be made by scientists to the foregoing argument. Granted, they will say, that the process of verification which you describe is a reflective one, still reflective processes are just as liable to error as direct; the question is, what is the criterion of truth in such processes, whereby it is distinguished from error. This criterion, they will urge, is the direct observa-tion contained in them, and nothing else. Not there-fore the fact that the process is a reflective one, but the fact that direct observation is brought to bear upon theories and imaginations, is the important fact. It is of no consequence, except as matter of idle curiosity, that the process in which this test is brought to bear may be styled a reflective process.

Putting aside a possible ambiguity which might sometimes lurk in the word *direct*, in arguments of this kind, and supposing the word to be used, not for

[1] Problems, &c. Vol. I. p. 380.

Book I.
Ch. II.
───
§ 4.
Its importance
as sole source
of verification.

immediate observation, but in the technical sense in which I have used it throughout, I should reply as follows. Direct observation, you say, is the important fact. Important, I ask, in respect of what? It is important in respect of the particular conclusion reached, in each case of the reasoning being applied; it is important if we regard the reasoning as an individual instance. But if we regard the validity of the reasoning as a general process, or in respect of the conclusive power which it has in all cases alike, then the circumstance that it is a process of reflection becomes equally important with the circumstance that it brings direct observation to bear upon imagination. For the fact that a thing is what it is known as, or the inseparability of the objective and subjective aspects of a thing, which is the principle of reflection, is the final principle which may be exhibited as a common major premiss in case of need, which is common to all the several cases of verification.

I do not say that this circumstance is more important than, but of equal importance with, the circumstance of direct observation being brought to bear. Both are equally important in considering the validity of the process of verification as a general process. The attempt to give exclusive importance to the latter circumstance is but a more subtil form of the above noted prejudice of scientists in favour of science. The scientist mind asks what we can prove, busying itself with the objects alone, and neglecting the general question of the theory of proof, which turns upon the nature of the relation between the objective and subjective aspects. This narrow judgment, as it seems to me, needs correcting by introducing the more general and deeper question of the

BOOK I.
CH. II.
___
§ 4.
Its importance
as sole source
of verification.

connection of knowledge with fact. And this mode of viewing the case is the only mode in which the other problems of philosophy can be finally solved. For their settlement depends upon bringing into view the nature of the conclusive power which reasoning in general, as a process of thought, possesses. The theory of proof depends upon the phenomena of reflection, and therefore is a part of analytical philosophy, or as I call it metaphysic, and not a part of non-philosophical science.

Herein lies also the answer to another objection which I cannot but anticipate. You say, it will be urged, that Mr. Lewes *practises*, as Locke and others whom you call psychologists, as well as all men even in the physical sciences, practise, and always have practised and applied, reflection. Is not this the essentially important thing? Why demand that they should *recognise* it as their method, as well as practise it? The answer is, Because without recognition it cannot be made the most of. It is the recognition of the return to unity involved in reflection which enables us to use reflection as a final court of appeal, in deciding philosophical questions; which in fact develops reflection into a reflective method. The conscious application of the reflective process constitutes the reflective method, the method of philosophy as distinguished from the method of science.

It is this implication of reflection with the processes of science which makes it difficult to disentangle philosophy from science, which makes it possible for Mr. Lewes to represent the philosophical process as a scientific one. Reflection *per se* is common to both. It is only when we again reflect upon the process of science, that we distinguish the part played by reflec-

BOOK I.
CH. II.

§ 4.
Its importance
as sole source
of verification.

tion in it; and then only do we find ourselves in possession of the ultimate principle of verification, and become aware of what that principle consists in. For we have in the distinction of objective and subjective aspects, inseparable, and equated with each other, a perception which embraces all possible cases or kinds of objects, at the same time that it is a final perception in each case. No more perfect agreement can exist than that between the subjective and objective aspects of the same thing. It is at once the key stone of the connection of objects in reasoning, and the key stone of the method which connects them. But to be used in this way and for these purposes, it must be recognised for what it is, a method of methods, criticising all by the application of its one distinction of subjective and objective aspects.

The implication of a higher and more complex mode of consciousness with lower modes is a general fact which must never be lost sight of. The higher is not exclusive of the lower, but a modification of it, a new mode superinduced upon it. Reflection is not exclusive of direct consciousness, but, having been superinduced on primary, is inclusive of direct. And similarly, or rather consequently, philosophy, which is the application of reflection, is inclusive of science which is the application of direct consciousness. If this has been forgotten or not perceived by scientists, it has been equally misconceived by those philosophers who have turned reflection into a mode of thought directly furnishing us with special kinds of intuitions. The theory of an *intellectual intuition* of this sort is not fully combated, unless you can show on what basis of fact the mistake arose. It is not finally conclusive to argue, as Mr. Lewes does, that

BOOK I.
CH. II.

§ 4.
Its importance
as sole source
of verification.

we have no grounds of experience for the existence of such a faculty; for the faculty is only imagined because of some experience which is thought to be explicable in no other way than by supposing it. You must attack the alleged experience itself. You must analyse the experience, and having done so, you may then be able to show that its features are traceable to known functions of consciousness.

Intellectual Intuition is in this way traceable to two things, and explicable as a misinterpretation of them; first, to the fact that reflection has percepts (as well as concepts), and secondly, to the fact of subjectivity in reflection. These facts are perfectly compatible with the theory of reflection here maintained, with the validity of reflection and its derivative position with regard to primary consciousness, its controlling position with regard to direct; with its function as the basis of philosophy, and with its limitation to the possible phenomena of science. But it is not extraordinary that these facts should have been differently interpreted by minds subject to the strong reaction against the accidentalism, if I may call it so, of the eighteenth century, and dissatisfied with the bulwark which the Critical Philosophy professed to raise against it. A strong bias, distorting the facts of reflective consciousness,—that is the history not only of *intellectual intuition*, but of the whole of that philosophical movement of which it was one feature. The only reply to it is to be found in a more searching and, if possible, an undistorted analysis of the facts of consciousness.

But the theory of intellectual intuition has by no means the importance which Mr. Lewes attributes to it. He seems to have forgotten that the theory, as

Book I.
Ch. II.
—
§ 4.
Its importance
as sole source
of verification.

held by Schelling and others, was definitively abandoned by Hegel, the final exponent of the post-Kantian method of idolising the faculty of reflection, who replaced percepts by concepts, and the process of perceiving by that of conceiving, with its single law of Contradiction. It was in the *Phänomenologie des Geistes* that this substitution was made; and we may see from the preliminary announcement of the scope of that work, at the very beginning of its long Preface, that the abandonment of intellectual intuition was a cardinal point of the whole. "If the true exists only in, or rather as, that which is named now intuition, now immediate knowledge of the absolute, religion, being,—not in the centre of the divine love but the being of that centre itself,—then from that point of view, we should require not the form of the concept but its very opposite, in order to expound philosophy. The Absolute is then proposed not to be conceived (*begriffen*) but felt and intuited ; not the concept but the feeling and intuition of it are to be its spokesmen and to be expressed."[1] Hegel then, who builds solely with the "form of the concept," is untouched by Mr. Lewes' objection, that a special faculty, such as intellectual intuition, is required for acquainting us with the Supra-sensible, since he breaks with all such faculties, and yet constructs, if not what he calls a supra-sensible world, yet one which includes all that Mr. Lewes calls the supra-sensible.

Now it is not only against intellectual intuition, but also against Hegel's construction which dispenses with it, that the conception of reflective consciousness, for which I contend, proves its validity. The

[1] Werke, Vol. II. p. 7.

BOOK I.
CH. II.

§ 4.
Its importance
as sole source
of verification.

doctrine that reflection is *derivative* and *analytic* (as well as subjective),—derived ultimately from primary perception and analytic of percepts,—brings us back to the real and actual world, the world of science, and prevents those flights which are the consequence of taking the bare logical principle of contradiction as the source of existence.

Let us hear one of the most recent disciples of Hegel on this point, Mr. Green, whose General Introduction to Hume I had occasion to appeal to in the foregoing Chapter. "That which happens, whether we reckon it an inward or an outward, a physical or a psychical event,—and nothing but an event can, properly speaking, be observed,—is as such in time. But the presence of consciousness to itself, though as the true 'punctum stans' (Locke, Essay II. Chap. XVII. Sec. 16) it is the condition of the observation of events in time, is not such an event itself. In the ordinary and proper sense of 'fact,' it is not a fact at all, nor yet a possible abstraction from facts."[1] The presence of consciousness to itself *not an event in time*, but the condition of observation of such events. Here is the conception to which I oppose the doctrine that self-consciousness is derivative and analytic. The opposition of this to the Hegelian doctrine brings out its true significance, the true significance of them both. But Hegel's doctrine is not met by a denial of intellectual intuition.

How it stands with the presence of consciousness to itself, will be better seen when we come to the analysis of elements in primary consciousness. Meantime, an Hegelian might ask me, why, seeing that reflection is a derivative of primary consciousness,

[1] General Introduction, § 142. (Vol. I. p. 121).

Book I.
Ch. II.

§ 4.
Its importance
as sole source
of verification.

and conception a derivative of perception. I deny of
the one derivative what I claim for the other, the
character or function of giving ultimate truth? why
I claim this character for reflection, while denying it
to conception? My answer is plain. In the first
place, this character is claimed, not for reflection in
the rough, but for reflective perception, the union of
the reduplicative mode of consciousness with perception its primary function. In union they are a mode
of final analysis. In the next place, there is a great
difference between reflection in the rough and conception, notwithstanding that both are derivative.
Conception as a process *per se*, the movement in conception taken alone, does not analyse, but combines.
It leads back to percepts, and ends as it began with
them; with percepts corrected and enriched, but still
with percepts. It leads back to a primary of the
same kind as that from which it was derived. Reflection on the contrary does not lead back to primary
consciousness, which was its original. Reflective and
direct consciousness replace and supersede their parent. Reflection, then, though derivative as conception is, yet in its union with perception is ultimate.

Hegelians of course will never allow that conception is derivative at all. The primary character
of conception is their fundamental position. And
since the *nexus* of phenomena is one of the chief
things to be accounted for in philosophy, or in other
words, since philosophy has to interpret in subjective
terms the actual connection between phenomena
which we perceive objectively; and it is admitted on
all hands that conception combines percepts; it follows that, if it cannot be shown that relations are
given in and with the minutest fragments of percepts,

Book I.
Ch. II.

§ 4.
Its importance
as sole source
of verification.

the *minima sensibilia*, then we must attribute the nexus of phenomena (subjectively speaking) wholly to conception, and then also we must hold, with Hegel, that conception is not derivative but primary.

But it can be shown. The minutest fragments of percepts, of perceived or represented sensations, occupy either time or time and space together; and these time and space elements, being common to all such fragments, connect them together and are the common bond of union between them. They are the *continua*, of which the differences of feeling are the *discreteness*. Not the logical law of Contradiction, under which percepts are connected and concepts formed, but the perceptual elements of Time and Space, are the nexus which makes the world organic. Be this however as it may, (and the present Chapter is not the place to redargue the basis of Hegelianism), the question is at any rate one which must be discussed before the tribunal of Reflection, or in other words, as a question of reflective and not of direct consciousness. It is moreover a question of *analysis*. In vain will Hegelians appeal to reflective analysis to destroy Locke, Berkley, and Hume, and reject its authority when appealed to against Hegel. Reflective analysis is the true Dialectic.

Book I.
Ch. III.

§ 1.
Self-contradic-
tory nature of
Things-in-
themselves.

CHAPTER III

§ 1. To the days of Jacobi, Fichte, and Schelling, belongs, as Herbart[1] tells us, the complet :

> Da die Metaphysik vor Kurzem unbeerbt abging,
> Werden die Dinge-an-sich jetzo *sub hasta* verkauft.

Which may be rendered, duly preserving the elegant smoothness of the original :

> *Oyez!* Things-in-themselves to be sold by public auction !
> Since Metaphysic is dead ;—dead without leaving an heir.

Poetry is catching, and it is but natural that such an effort as the above should call forth an equally poetical reply. Here it is :

> What though Things-in-themselves have been dispersed by an
> auction ;—
> Who was the auctioneer?—Why, Metaphysic herself.

The meaning and the justice of this reply will immediately become apparent, for the fact is that this whole question lies in a nutshell. The application of the distinction, explained in the foregoing Chapter, between direct and reflective consciousness, clears up the whole position, and in this way. When

[1] Allgemeine Metaphysik, § 94. Werke, Vol. III. p. 261.

THINGS-IN-THEMSELVES AND PHENOMENA.   163

Book I.
Ch. III.

§ 1.
Self-contradic-
tory nature of
Things-in-
themselves.

we approach external phenomena by the direct
method, asking what and how much we can make
out about them, we come at last upon a residuum
about which nothing can be made out, we reach the
limit of our possible knowledge by the means at our
disposal. This residuum is the supposed Thing-in-
itself, while all that we can make out about the
phenomena is knowledge, is phenomenal. If our
direct enquiry has been statical, that is, into the
analysis of the phenomena themselves, the residuum
appears in the shape of a substance, or substratum,
of phenomenal attributes, qualities, or properties. If
the enquiry has been dynamical, that is, into the
causes, conditions, or genesis, of the phenomena,
then the residuum appears in the shape of an unknowable First Cause. But in either case the residuum is found *because of our own assumption;* we
have put it into the phenomena by adopting the
direct method of enquiry.

The same is the case if we take internal phenomena, states of consciousness, as our objects, and
approach them by the direct method, which is the
method of psychology. If the enquiry is statical, we
come to a residuum in the shape of a substance or
substratum of states of consciousness, we come to
what is called a Soul, or Mind, or Spirit. If the
enquiry is dynamical, we come to the residuum in
the shape of a power, an agent, that is, a Self or an
Ego. Again the residuum is found in the phenomena
solely in virtue of our assumption in choosing the
direct method of enquiry.

The difference of the object-matter, external phenomena in the one case, internal in the other, makes
no difference to the result in question, which depends

Book I.
Cн. III.

§ 1.
Self-contradic-
tory nature of
Things-in-
themselves.
upon the adoption of the direct method. For if we
enlarge the field of enquiry by putting together the
two former fields, and taking internal and external
phenomena together, the result is just the same, so
long as we do not change the method. We then get,
on the statical line of enquiry, to a residuum which
is the supposed *absolute* substance underlying both
orders of phenomena, *e.g.* Spinoza's Substantia with
its attributes, the two known to us being extension
and thought; or we get, on the dynamical line, to a
supposed Force or Power, evolving eternally, but
not exhausting itself by evolving, both orders of
phenomena at once.

Things-in-themselves, then, are the necessary result of enquiries pursued on this method, because the
adoption of the method involves the assumption of
an existence separable from and independent of our
knowledge of it. We only find in it at last what we
imported into it at first. We choose to adopt the
attitude of enquiry into a ready made Existence, and
then are surprised to find, when we have discovered
all we can about it, that the ready made existence
still remains partially unapproachable, still continues
to be a ready made existence; are surprised too by
the logical contradiction in which we are entangled,
namely, that of affirming an Existence about which
we have no knowledge.

"Before they were aware, he led them both
within the compass of a net," the *Flatterer* being,
in this case, the flattering delusion that philosophy
can be mastered by the methods of science. Unaware that philosophy has a method of its own, a
method elaborated by the toil of many generations
of philosophers, the eager scientists, deluded by the

THINGS-IN-THEMSELVES AND PHENOMENA.   165

BOOK I.
CH. III.

§ 1.
Self-contradic-
tory nature of
Things-in-
themselves.

success of their own methods within their own domains, expect to enter on the heritage of philosophy,
and with their methods to reap the fruits where
another method has digged and planted and watered.
But before they are aware they find themselves en-
closed within the compass of a net; one and all, one
after another, in they march;

> " Mixta senum ac juvenum densentur funera. Nullum
> Sæva caput Proserpina fugit."

One insoluble contradiction awaits them all, one fate,
one absurdity, the ancient puzzle of the Thing-in-
itself, which in one and the same breath they must
affirm to be unknowable and affirm to be known to
exist. In vain they toss and struggle. Their method
is able to get them in, but powerless to get them out.
There is nothing for it but to await their extinction
and abuse philosophy.

But now let us turn to the simple solution of the
puzzle by the method of reflective consciousness, the
method of philosophy. This method consists, as the
foregoing Chapter has shown, in asking what is the
meaning or connotation of terms in their relation to
consciousness alone. Simply applying this to the
term *Existence*, the answer is obtained in quite general terms, Existence means presence in consciousness; *esse* means *percipi*. Now if that is the meaning
of the term existence, it is clear, first, that there cannot be an unknowable existence, and secondly, that
we cannot frame the notion of such an existence,
because the two terms which compose the notion destroy one another; when you have fixed one, the
other has gone. The contradiction between the two
terms applies to the notion equally when taken ob-

Book I.
Ch. III.
―――
§ 1.
Self-contradic-
tory nature of
Things-in-
themselves.

jectively as when taken subjectively, equally to existence as thing, and to existence as thought. Accordingly, in the phrase *unknowable existence*, one of the
terms is a word without a meaning, a word not standing for a thought or a thing.   •

But what comes of the application of this result
to the former one, to the conclusion reached by the
direct method, to the existence there discovered beyond the limits of possible knowledge? True, it may
be said, the reflective method convicts the conclusion
reached by the direct method of involving a contradiction; we grant so much; but the question is, can
it go beyond this, can it show what is the *truth* as
well as what is the error in that conclusion? I reply
that it can. The unknowable existence affirmed by
the direct method is not unknowable strictly; it is
only supposed to be unknowable so long as we forget or neglect the reflective result, the meaning of
the term *existence*. There is an existence unknowable in certain most important respects, in *all* respects
which *science* busies itself with; that residuum is a
scientifically unknowable; but it is not a philosophically unknowable, since this residuum being known
to exist is known to be possibly present to consciousness, a *percipiendum* as well as an existent. Science
merely goes *too far* in its assertion that there is beyond science an unknowable existence; there is such
an unknowable existence for science, but not for philosophy: the reflective method which equates the
objective and subjective aspects goes beyond the
limits of science, goes *as far as existence goes*. Observe too that, in saying there is an unknowable
existence *for science*, we limit and restrict the term
unknowable existence; it is then no longer strictly

THINGS-IN-THEMSELVES AND PHENOMENA.   167

Book I.
Ch. III.

§ 1.
Self-contradic-
tory nature of
Things-in-
themselves.

unknowable. Science goes too far in affirming it, and in doing so is wrong in point of fact. It might have gone too far in assertion, that is, gone beyond its legitimate evidence, and yet have been right in what it asserted. But it is wrong as a matter of fact. There is no unknowable existence. The method of reflection equates existence with perception.

There are accordingly two regions of thought and existence, one which science and philosophy have in common, the other which philosophy going beyond science occupies alone; and both regions are alike phenomenal. However little we may know of the supra-scientific region, we are not in a region of Things-in-themselves. *Thing-in-itself* is a word, a phrase, without meaning, *flatus vocis*. It is only for science that it is an illusion, an existent opposed to phenomena. True, we may people the supra-scientific region with phantasms, we may falsely and groundlessly imagine all sorts of beings in those quarters, but they will all be phenomena, as much phenomena as are the immediate presentations to sense, as much phenomena as if they lived by bread. Not only, then, is the Thing-in-itself abolished by reflection, but the distinction between Thing-in-itself and Phenomenon is abolished with it. Everything is phenomenal; everything is relative; in the particular sense of relativity, that of having a subjective in relation to an objective aspect. The question so often proposed—What is the Thing-in-itself, is really a non-sense. It *intends*, no doubt, to ask (and the question is perfectly legitimate) what is the nature of this or that existence in the supra-scientific but intra-philosophic region? If so meant and so understood, it may often be an useful and important ques-

Book I.
Ch. III.
———
§ 1.
Self-contradic-
tory nature of
Things-in-
themselves.

tion. But supremely important is it to remember that Things-in-themselves are a misnomer, words without meaning. The proper definition of a Thing-in-itself is:—a thing known to exist without any knowledge of *what it is* being possible: which definition sufficiently shows the self-contradiction involved in the term.

I turn now to a feature of the question which is of the greatest importance, the perfect *generality* of the result arrived at. Although an individual reasoner in no way transcends his own individual consciousness, yet he reaches a result which is true of all phenomena and under all relations; that is to say, he reaches a result which is true of phenomena *qua* phenomena, whether they are phenomena in presentation or in representation, in perception or in conception, in memory or in imagination, and whether they are objects to the individual's own consciousness or to the consciousness of other individuals. The truth reached is this, that phenomena *qua* phenomena are the whole of existence, and at the same time the whole of consciousness. There is no Thing-in-itself beyond them. And the term *phenomenon* is a term expressing at once the double aspect, subjective and objective: the subjective aspect of existence, the objective aspect of states of consciousness.

The term *Noumenon*, which is employed to denote Thing-in-itself as opposed to phenomenon, by connoting its merely conceptual character as opposed to its perceptual, bears witness itself against the doctrine which it is used to expound. For it introduces a distinction into consciousness, and while discarding the objects of one function retains those of another, —noumena, the objects of conception. It retains

Book I.
Ch. III.
——
§ 1.
Self-contradic-
tory nature of
Things-in-
themselves.

these objects in consciousness generally, excluding them from perceptive consciousness; they are still phenomena, because they are still objects with a subjective aspect.

Objects of representation again, as opposed to objects of presentation, are still phenomena in the same sense; so also are objects of memory; so also are objects of imagination. In fact objects of memory and of imagination are represented objects; memory and imagination being determinate modes of representation. It is not presentations only, that is, immediate feelings while actually felt, that are phenomena.

In short the term *phenomena* is perfectly general; it has a connotation, *union of the two aspects*, and whatever answers to this connotation, no matter what may be the particular mode in which it answers to it, is included under and is denotable by the term. The *accidents* of the object so defined,—phenomenon, the union of the objective and subjective aspects,—an object be it noted of reflective consciousness, inasmuch as reflection discovers the connotation of the term denoting it, the object unanalysed being given by primary consciousness,—the accidents may be here disregarded. We are not now occupied with any of the modes of phenomena, but with phenomena themselves merely as such.

But it may possibly be thought that, when we take account of the other distinction mentioned above, namely, phenomena as objects to the individual reasoner's own consciousness and phenomena as objects to the consciousness of other individuals, we cannot maintain the result here reached in its generality, without transcending the individual's own consciousness; or else that, on the other hand, if we

Book I.
Ch. III.

§ 1.
Self-contradic-
tory nature of
Things-in-
themselves.

refuse to make a statement which transcends this consciousness, then, since we limit our statement about phenomena to one class of them, namely, the phenomena of our individual mind, we must renounce the claim of perfect generality for the result. For instance, Professor Clifford writes:[1] "Herein is a distinction between Self and Not-self, which is far deeper than that between Subject and Object; the distinction between *You* and *Me*. My conception of your consciousness is a symbol that differs from all objective symbols in this ; that they can be expressed in terms of my feeling: this never can. In the primary sense of the word object, then, I know directly one object, this table, which is in my consciousness; I infer symbolically a number of similar objects in the consciousness of other men. Out of all these there now arises a still more symbolic conception ; the table, as object in the consciousness of man."

In what sense the distinction between the consciousnesses of two individuals is "deeper" than that between Subject and Object, is not made clear in this passage. Let us see what it really is. The former is a distinction between separables, the latter between inseparables. The former is a cleft, the latter a bridge over it. In this sense no doubt the former is the deeper of the two. But it does not go deeper into the analysis of the phenomena. A distinction which bridges a cleft must in analysis be deeper than the separation which makes the cleft, for it brings to light a circumstance of sameness where only difference had been observed before; it furnishes an answer, where the other had proposed a question.

[1] *The Academy*, for Feb. 7, 1874.

Book I.
Ch. III.

§ 1.
Self-contradic-
tory nature of
Things-in-
themselves.

Professor Clifford in this passage takes the Subject to mean the individual Subject, and the Object
to mean (1) the object of that single individual, and
(2) the object of human beings collectively. And
then he is perfectly right in holding that this second
object is doubly symbolical, as he calls it. First I
have to get *your* table, and then *mankind's* table. But
I must begin with *my own* imperfect and rudimentary
table. That is the first step of all; and that furnishes
me with my conception of *your* rudimentary table,
and finally of *mankind's* perfect one, which is Professor Clifford's Object in the second sense. Without
my table to begin with, I should be *totally* unable to
reach yours and then mankind's; *totally* unable to
get across the "deeper distinction" between You and
Me; your table, and consequently you, would to me
be non-existents. At the same time, when by aid of
my table and comparison combined I have reached
mankind's table, then mankind's table becomes my
table and my object, just as much as my merely
individual table was originally. The distinction between Subject and Object as I draw it, in the individual, survives, as it originated, the whole process;
is last as well as first in it, and includes under it the
steps from my object to yours, and from yours to
mankind's.

It is, therefore, a mistake to suppose that a greater
difference exists between my consciousness and yours
than can be bridged by, that is, shown to be a case of,
and so subsumed under, the distinction between object and subject in a single, say my, consciousness.
For it is not the fact that "my conception of your
consciousness is a symbol that cannot be expressed
in terms of my feeling." This would be true only if

BOOK I.
Ch III.

§ 1.
Self-contradic-
tory nature of
Things-in-
themselves.

we took "my feeling" to mean my *presentative* feeling. To you that consciousness is presentative which to me is representative; that is, I represent to myself what *I imagine* you to be feeling presentatively. And I express it to myself as so represented in terms of my feeling, in terms which derive their significance solely from feelings which I have myself presentatively experienced. True, I cannot have your presentations *as subjective presentations* of my own, but I can have them as representations of the imagination, which is to have them as my objects.

On the other hand, it is not the fact that *all* "objective symbols" can be expressed in terms of my feeling, unless we take "my feeling" to include representations. There are some objects which cannot be presentations either to you or to me, but are inferences only, representations of imagination ; the atoms, or shall we say vortices, of the ether, for instance, or the far side of the moon. The change in physical conditions which would be requisite to enable me to have your feelings as presentations of my own, say a perfect nerve communication between my brain and yours, is not surely a much more violent hypothesis than would be the hypothesis of a change which should enable me to see the far-side of the moon or the atoms or vortices of ether. The impossibility of *my* having *your* feelings does not affect their nature as phenomenal in point of *kind*, but merely the mode of their phenomenality to me.

Professor Clifford's doctrine, then involves taking the term "my feeling" in two different senses at once. For in order to include all objective symbols under what can be expressed in terms of "my feeling," the term must be understood in the large inclusive sense

Book I.
Ch. III.

§ 1.
Self-contradic-
tory nature of
Things-in-
themselves.

which is the proper one; and in order to exclude "my conception of your consciousness" from the same category, the same term must be understood in a narrow and improper sense.

We are in this dilemma,—either an individual Subject has some knowledge of other people's feelings, and then they are part of his object, and the same in kind, *qua* phenomena, with his own, or else he has no knowledge of them at all, that is, they are non-existent to him. The former alternative is confessedly the true one. Therefore, however imperfect the Subject's knowledge of other people's feelings may be, he knows at least this; that they are the same in general kind as his own, namely, phenomenal. If assumed at all, they are phenomenal to him; and again, if assumed at all, they are assumed as phenomenal to the immediate Subjects of them. In both ways they are phenomenal in point of kind.

The distinction, therefore, between *my* feelings and *yours* can never be a foundation or basic distinction in metaphysic, which treats of the *nature* of things, because it is not a difference of *kind*, but is merely one of time, place, and circumstance. It is a distinction in science, the science of psychology, presupposing at least one individual Subject already existing, separate from other individual Subjects real or supposed. The separation of individual Subjects is the first step in psychology; and here no doubt the separation between conscious individuals, conscious centres of presentation, is a most important fact. For, as I have shown in the foregoing Chapter, verification depends on presentation repeated in reflection, and therefore, since presentation is always individual, verification is always an individual process.

Book I.
Ch. III.
___
§ 1.
Self-contradic-
tory nature of
Things-in-
themselves.

The same circumstance also in ethical matters determines the supremacy of conscience. The convergence and agreement of individuals is the discovery of the single true world among the many real ones. We do not need the supposition of a single world-consciousness corresponding to the single true world. For, since the presentative differences of conscious individuals can be bridged in representation, the harmony of all individuals is equivalent to a single individual;—much as if all clocks in the world were right by one another and the sun, which they might conceivably be, without a world-clock other than the world itself.

Quite recently Professor Clifford has published a paper[1] in which his views on this subject are drawn out more fully. He there coins the word *eject* to mean "your feelings," as opposed to *object* or *phenomenon* which means "my feelings." "These inferred existences," he says, "are in the very act of inference *thrown out* of my consciousness, recognised as outside out of it, as *not* being a part of me."[2] And an object of the consciousness of mankind collectively, *e.g.* "this table, as an object in the minds of men," is distinguished by the title of *social object*. There are three things to be distinguished, the individual object, the social object, and the eject. And of his instance of the social object he says, "This conception symbolises an indefinite number of ejects, together with one object which the conception of each eject more or less resembles."[3]

These words will serve to reveal what I think is

[1] On the Nature of Things-in-themselves. In *Mind*, No. IX. p. 57, Jan. 1878.

[2] The same, p. 58.　　　　　　[3] The same, p. 59.

Book I.
Ch. III.
——
§ 1.
Self-contradic-
tory nature of
Things-in-.
themselves.

the lurking fallacy. It is not the case that a conception like "this table as an object in the minds of
men" symbolises *ejects*; it symbolises the speaker's
representation of them. And if it should be replied,
that this is provided for by the words "together with
one object, &c.," then I answer, that a mere *together* is
not enough; the object in the speaker's own mind,
the "individual object," supplies him with the whole
interpretation of what those ejects are, as well as
with his whole grounds for assuming their existence. He knows nothing whatever of the ejects,
except as part of some "individual object," or objects, of his own.

But it will be said, the ejects *exist* all the same,
and if they did not exist, the individual object would
never be completed or moulded into the social object. Very true, they exist, and the signs they give
of their existence are what enable the individual to
develop his individual object into a social one. But
this is entirely irrelevant to the question of *what* the
ejects are; it is not a point of metaphysic but of
psychology; affects the history of individual and
social development, but does not touch the *nature* of
consciousness, whether in this individual or in that.
If the ejects exist, still they are in their nature phenomenal. That *I* can only *infer* their existence,
makes no difference in this respect. If I do infer
them, they are phenomenal to me. If I do not infer
them, still they are phenomenal to you.

But what inextricable confusion is introduced
into philosophy by thus mixing up the psychological
point of view with the metaphysical. The question
of *nature*, of kind, of the τί ἐστιν, of a thing, is shelved
and replaced by a question as to whence it comes and

BOOK I.
CH. III.
———
§ 1.
Self-contradic-
tory nature of
Things-in-
themselves.

whom it belongs to: and then a difference in respect of this incidental circumstance, namely, whether a feeling is *mine* or *yours*, is made the ground of inferring difference and even opposition of *kind* between them.

For what can be more opposed in kind than Things-in-themselves and Phenomena? Now it is the main drift of this paper of Professor Clifford's to show, that what he calls *ejects*, (opposed it will be remembered to *objects*, which are phenomena), are Things-in-themselves. "Consciousness," he says, meaning I suppose *other people's*, "is a complex of ejective facts,—of elementary feelings, or rather of those remoter elements which cannot even be felt, but of which the simplest feeling is built up."[1] Presently we read, "Hence a feeling (or an eject-element) is *Ding-an-sich*, an absolute, whose existence is not relative to anything else. *Sentitur* is all that can be said."[2] These elements of feeling are then called *Mind-stuff*. "A moving molecule of inorganic matter" [which is an object, a phenomenon, interpretable in terms of '*my* feeling'] "does not possess mind or consciousness; but it possesses a small piece of mind-stuff."[2] The elementary feeling, then, which is *Ding-an-sich* is not matter. Neither is it consciousness;—unless it be (shall we say?) an unconscious variety. I should like too to hear how it differs from *spirit*. However, I am not here concerned with the theory of Mind-stuff. For aught I know, there may be many other objects of consciousness besides matter. My wonder is to find any one ambitious of

[1] On the Nature of Things-in-themselves. In *Mind*, No. IX. p. 64. Jan. 1878.

[2] The same, p. 65.

having Things-in-themselves as an item in his philosophical system. Mind-stuff may possibly be a fact; but if so, it must certainly be phenomenal and stand in relation to other facts. The reason for which is, that the condition precedent, or origin, of things, conceive it as you will, is a question of science, whether physical or psychological, and is subordinate to the question of nature, which is the question of metaphysic,—*what* that condition is conceived as being. For we cannot know *that* anything exists, without having some knowledge of the thing itself, that is, of some of its attributes or characteristics. Otherwise we should have to conceive it as *an existent Nothing.* And this is precisely what a Thing-in-itself is.

§ 2. What has now been said might appear sufficient to bring the present discussion to a conclusion, since the distinction between Things-in-themselves and Phenomena has been fully exhibited. The whole question lies, as I said, in a nutshell. But it will perhaps be well, seeing the embarrassment which the point occasions to philosophising scientists, and the importance which it has assumed in the eyes of a demi-semi-Kantian public, to say something of the shape which the Thing-in-itself assumed in Kant's hands, and on the subsequent fate of that theory of it, of which Kant was the author.

The *locus classicus* on the point is that chapter in the *Critic of Pure Reason* entitled "On the ground of distinguishing Objects generally into Phenomena and Noumena." It is the locus classicus because it gave precision to the notion of Things-in-themselves, by basing it upon a theory of the nature and laws of consciousness. It demonstrated, from that theory, that the Thing-in-itself, which before had been a mere Unattainable, had a necessary existence, at the same time that it was necessarily unknowable. The Thing-in-itself before Kant is very different from the Thing-in-itself after Kant. It developed into a nuisance. Prior to Kant we meet with it continually. Aristotle has ὑποκείμενα ἄνευ αἰσθήσεως, ἃ ποιεῖ τὴν αἴσθησιν.[1] Still it was not definitely presented as something completely and entirely unknowable. A sort of half unknowability is the shape in which it appears, in those instances which Sir W. Hamilton quotes as Philosophical Testimonies to the Limitation of our Knowledge;[2] including indeed Kant among the rest, yet apparently without a suspicion of Kant's difference from all his predecessors on this point. For side by side with the unknowable, prior to Kant, we meet also with expressions of the adequacy of knowledge to existence; adequacy, that is, in some respects. For instance, in Aquinas we have the following: "Est enim proprium objectum intellectus ens intelligibile, quod quidem comprehendit omnes differentias et species entis possibiles: quidquid enim esse potest intelligi potest."[3] And to come to Kant's immediate predecessor, Wolff, we find him saying: "Ens omne, quâ tale, cognoscibile."[4]

But Kant it was who put the old notion of the Thing-in-itself on a new footing; and, since he

---

[1] Metaph. Γ. v. p. 1010. b. 32.
[2] Discussions, App. B. pp. 634 649. 2nd edit.
[3] Summa Phil. Contra Gentiles, Lib. II. cap. 98.
[4] Logica, § 28.

deduced it from his theory, therefore the counter demonstration of the non-existence of the Thing-in-itself involved the fall of his theory. This in fact was the point at which the expansive force of the method of reflection, admitted by Kant under the name of Apperception, shattered the fabric of the Critic of Pure Reason, as was noticed in Chapter I.

But now let us see what Kant's doctrine really is. The main points of the Chapter mentioned above[1] may be given briefly as follows:

The *transcendental* use of a Concept (under any of the Categories or principles of Synthesis) is in applying it to Things *generally* and *in themselves* (p. 226).

Ontology lays claim to give a doctrine of things generally (p. 231).

It must give place to an *Analytic* of the Pure Understanding (*id. id.*).

The pure categories, taken alone, have a transcendental significance, but no transcendental applicability (p. 231-232).

In the mere consideration that it is requisite to an Object that it be in relation to our *Intuitive* faculties, as well as to our Understanding, lies the distinction between Phenomena and Noumena, and the necessary assumption of the *existence* of the latter as correlates of the former (p. 234). [Here we have the necessity of the existence of Noumena, founded on the duality of the faculties, intuition and understanding.]

From this follows naturally the mistake, that, since we can form the concept of an Object generally, (i.e. not limited to be in relation to our intuitive

---

[1] "*On the ground, &c.*" Kritik d. R. V. pp. 224-240. Hartenstein's edit. 1853, which is from the Second Edition of the Critic.

faculties), the same categories which helped us to form this concept must help us also to *determine* it, and so to *think* the Noumenon as such and such in reality and in itself (p. 235). [Here we have the non-necessary but natural mistake about the nature of the Noumena.]

In consequence of this distinction we have Noumena distinguished into:

1. Things so far as they are *not* objects of our sensible intuition;—Noumena in a negative sense.

2. Things so far as they *are* objects of a non-sensible intuition;—Noumena in a positive sense (p. 235-236).

Since we have no such cognitive faculty at all as Intellectual Intuition, Noumena for us must always mean Noumena in the negative sense (p. 236).

Problematic Concepts are such as, though containing no contradiction, and even hanging together with other cognitions and being conterminous with given concepts, yet cannot have their objective reality shown. A Noumenon, always in the negative sense, is an instance of such a Limit-Concept, and Problematic Concept, and thus is of merely negative use. Yet it is not arbitrarily feigned, but is brought in by the limitation of our sensibility, without adding anything positive beyond the range of sensibility (p. 237).

The Noumena here defined are very different from the *mundus intelligibilis*, the world as thought or reasoned, in other words, the system of laws of nature. The distinction of *mundus sensibilis* and *intelligibilis* does not belong to the system of the Critic, and leaves

THINGS-IN-THEMSELVES AND PHENOMENA. 181

Book I.
Ch. III.
———
§ 2.
Kant's theory
of them.

the question of Noumena quite untouched. That question is, Whether there are Noumena as legitimate objects of thought beyond the range of the *mundus intelligibilis;* and this question we have answered in the negative (p. 238-239).

————————

The above abstract from Kant's famous Chapter gives the whole theory of Things-in-themselves, as it exists for Kantians at the present day. We can see from it the supposed logical basis of the belief in those Things. The existence of two faculties, intuition and thought, or in modern and more philosophical terms, of two functions, perception and conception, of which one, conception, is held to be wider in its range than perception,—this relation between the two functions gives a reason to the belief in Noumena, which are the objects of conception in that part of its range which is larger than the range of perception. But the duality of function alone would not suffice to maintain the belief, nor yet I think would the inverted relation of the two functions, as to range, suffice to maintain it, if the process of reflection was duly generalised, and the supposed Noumena by this means included, not indeed in the field of a determined faculty of intuition or function of perception, but in the less determined field of objects of consciousness generally. Kant's Noumena in a negative sense are what I have called above noumena *per accidens,* noumena in a certain limited respect, unknowables in respect of perception. And this notion of noumena *per accidens* is given only by reflection.

There are, then, two ways in which we may endeavour to extricate ourselves from Kant's position. One is that which I have already taken. by generalising the process of reflection, and thus equating the objective and subjective aspects; which method leaves of course a number of respects in which objects may be unknown to us, while maintaining one respect at the least, in which all objects must be known to us, namely, as existing phenomenally, though without further determination of their mode of existence. Existence is, on this view. the object of a representation. For reflection is restricted to be reflection on primary states of consciousness. and these are exhaustively divided into presentation and representation.

The other way of extrication consists either in equating the two functions whose discrepancy gave rise to the supposed demonstration of noumena, or else in deducing one of them from the other. I am not acquainted with any system which has followed the first of these alternatives. But the second alternative divides again into two branches, according as perception is held to be a derivative of conception, or *vice versa*. In both of these two branches a fundamental unity of process in consciousness is obtained. But both branches rest upon an application of the method of reflection, and are in fact supplementary to the first way of extrication. They presuppose the consideration. that the source, the thinking-principle, which throws up the categories, so to speak, and works in and through them,—they being the creators of the Thing-in-itself,—must be adequate *subjectively* to that Thing-in-itself *objectively*, must be a correlate of it. But this source is the Apperception (empirical

or transcendental) indicated by the words *Ich denke*.[1]
Therefore the Thing-in-itself is not without some
adequate subjective correlate, that is, is *not* what it
was taken to be, a Thing-in-itself.

The first of the two supplementary branches is
represented by Hegel. His law or process of Identifying Contradictories, with the Concept as absolute
Basis,[2] which is described by Hegel as the Concept of
the Concept itself, *der Begriff des Begriffes*,[3] is the
representative and substitute for Kant's transcendental Apperception working in and by the Categories
of the Understanding; it is that single process, or
single force, which I have just described as throwing
up and then working in and by the categories. And
being concrete, a concept which throws up concepts,
as well as a law of their identification, it includes percepts as part of itself. Elsewhere I have criticised this
theory, and endeavoured to show that it is not compatible with a true analysis of consciousness. I will
now remark only upon its genealogy,—which may
be thus deduced. The Conception in concepts comes
directly from Kant's transcendental apperception, the
concepts from his categories. The categories are the
descendants of " innate ideas." These again spring
from the Realist interpretation of Aristotle's Universals; and these universals, so interpreted, are an
attenuation of the Platonic Ideas, which had a separate existence beyond and beside the things of which
they were predicable. If anything could have established the truth of the absolutist tradition, in the

---

[1] See Hegel, Die subjective Logik. Vom Begriff im Allgemeinen.
Werke, Vol. V. p. 14.

[2] The same, p. 5.

[3] The same, p. 12.

teeth of facts, it would have been the subtil genius of Hegel.

Of the second supplementary branch, that which· reverses Hegel's theory, and subsumes conception under perception, I shall not now speak. It is the theory which I hold to be the true one, and in future Chapters I shall devote considerable space to its consideration. Unity of function in consciousness is attained by both theories; both go to the extreme limit of consciousness and of existence; both leave enormous tracts of unknown existence within those limits; both hold Things-in-themselves to be *flatus vocis*. But Hegel's theory does not separate the analytic from the constructive branch of philosophy, and consequently does not distinguish between the science of the actual world and the philosophy of hypothetical worlds. Both branches are by him included, undistinguished, in one and the same philosophy. Hegel at once overrides science and gives the analysis and genesis of both worlds at one and the same time, by one and the same theory. It is in this, and not in its inclusion of the Thing-in-itself, that Hegel's pretensions are exorbitant. He makes philosophy remove its neighbour's landmark, by first making it overlook a landmark of its own, the distinction between analysis and genesis.

We can now see the position occupied by those various theories in the constructive branch of philosophy which were enumerated in Chapter I.[1] They may not be Ontological, that is, they may make no claim to override or obliterate the distinction of objective and subjective aspects; or to explain why there is existence, or why there is consciousness;

[1] Above, p. 77.

thus assuming that this distinction can be transcended. They may keep strictly and fairly within
phenomenal limits, while going to the full length of
the tether, the full extent of existence and of consciousness. They may, to put the thing in other
words, pitch upon some feature or features, in known
phenomena, as the central ones, as those which afford
an explanation of the rest, or from which the rest may
be deduced. No objection may lie against them on
the score of being transcendent or ontological. But
at the same time it must be remembered that, unless
such theories are framed with a distinct knowledge
and admission of the ground on which they stand, of
the province of philosophy to which they belong,
namely, to the constructive branch, and of the re-
quirements of that province, namely, that any theory
in it must be based on a previous metaphysical
analysis in the analytic branch,—they will not and
cannot be accepted as true hypotheses of the nature
of existence beyond the reach of science. It will be
said that this is to include all past theories in one
sweeping condemnation. I do not deny it. Time
has already condemned them as complete theories.
Time is only now bringing to light the conditions
upon which alone such theories can be legitimately
and securely framed. But this is very far from con-
demning those theories as *unprofitable*, or their authors
as dreamers. Without them where should we be?
Ubinam nescientiæ quæso? No, it is precisely in the
questions of the constructive branch of philosophy
that the sharpest stimulus lies to the study of philosophy at all. The *Why*, *Whence*, *Whither*, of this
transitory world, which is but a minute portion of
the world which is possibly actual, or actual to other

intelligences than ours, are the very questions in which we take the deepest interest. Let metaphysic once give us the *What* of this actual world of ours, and science its *How;*—then we shall be in a position to frame with advantage our first legitimate though imperfect solutions of its *Why, Whence, and Whither.*

§ 3. Having thus seen how philosophy deals with the question of Things-in-themselves, let us now see how it is sometimes dealt with by philosophical scientists. It is just one of those test questions which show a writer's capacity or incapacity for philosophy. It is so because its answer depends so directly and immediately upon the method of reflection. Grasp and apply that method, and you solve the question at once; ignore it or misconceive it, and all solutions are apparent only. This method, as we have seen, gives us things considered as existing in place of things as they exist; it gives us percepts instead of things independent of perception; things in their objective aspect instead of things in themselves. The Unknowable, or Thing-in-itself, becomes a word without a meaning; for all its meaning now belongs to things in their objective aspect, namely, to that portion of them which is a residuum after following up an analysis as far as possible on any particular line. This residuum is never an unknowable. It is the *continuation* of the known into the unknown. Its ultimate *laws* must be conceived as the same with those of the known. Hence it is no source of the *miraculous.* But call it unknowable, and it at once becomes *lawless*, the possible nay inevitable source

THINGS-IN-THEMSELVES AND PHENOMENA.   187

Book I.
Ch. III.

§ 3.
Scientist
theories of
them.

of miracles. Call it, however, what you will, you cannot say *anything* about it without making it relative, continuous with the known, and subject to some of the laws of the known. Therefore, that which was thought to be the unknowable, the thing-in-itself, becomes partially known, and belongs to the domain which I have called the constructive branch of philosophy.

Let us now see how Mr. Spencer deals with this question. Let us take his Chapter on Transfigured Realism.[1] The concluding paragraph runs as follows: "Finally, then, we resume this originally-provisional assumption but now verified truth. Once more we are brought round to the conclusion repeatedly reached by other routes, that behind all manifestations, inner and outer, there is a Power manifested. Here, as before, it has become clear that while the nature of this Power cannot be known—while we lack the faculty of framing even the dimmest conception of it, yet its universal presence is the absolute fact without which there can be no relative facts. Every feeling and thought being transitory—an entire life made up of such feelings and thoughts being also but transitory—nay the objects amid which life is passed, though less transitory, being severally in course of losing their individualities, quickly or slowly; we learn that the one thing permanent is the Unknowable Reality hidden under all these changing shapes."

Here we have what is Unknowable characterised as Power, as Reality, as an absolute fact, as universally present, as the one thing permanent. Five terms, which, since they are applied to what is un-

[1] Principles of Psychology, Vol. II. pp. 489-503. 2nd edit.

BOOK I.
CH. III.
——
§ 3.
Scientist
theories of
them.

knowable, must be words without meaning. Does
Mr. Spencer know the unknowable, or does he not?
Apparently it is a knowable unknowable that exists
for him.

This conclusion will not surprise us, when we
consider his description of the Realism which he
holds. since this description shows plainly enough
how he comes by it. "The Realism we are committed
to is one which simply asserts objective existence as
separate from, and independent of, subjective existence."[1] Simply asserts the direct denial of the fact
given in every exercise of reflection, which is the
non-separation of the objective and subjective aspects
of existence. Can we have a more evident proof
than this, of the necessity of adopting reflection as
the basis of philosophical method?

I have no doubt that what Mr. Spencer *intends* to
maintain by his doctrine of Transfigured Realism,
(since he cannot *intend* a contradiction), is perfectly
true and sound. When he says of the unknowable
reality, in the first cited passage, that "we lack the
faculty of framing even the dimmest conception of
it," he has in his mind *scientific* conceptions only, he
means such a concrete conception as could be directly
verifiable *in science;* he puts out of view the hypothetical. highly abstract, conceptions of philosophy.
But this only shows how inadequate to philosophy
scientific conceptions and scientific methods are, when
taken alone; how inevitably they lead to self-contradictions when employed to solve philosophical problems,—contradictions which are then supposed to
be inherent in the laws of consciousness itself; and
therefore how necessary it is to keep clearly drawn

[1] Principles of Psychology, Vol. II. p. 494. 2nd edit.

Book I.
Ch. III.
§ 3.
Scientist
theories of
them.

the distinction between the two domains and the two methods, of philosophy and of science. What Mr. Spencer *intends* by his Unknowable Reality is not, as the word *unknowable* suggests, Kant's Thing-in-itself; but, as partially knowable, it is the object of what I have called the constructive branch of philosophy, and corresponds to Mr. Lewes' Supra-sensible and Metempirical. This at least is the way I should find out of the contradictory notion which as we have seen Mr. Spencer has framed of it.[1]

Mr. Lewes' self-contradictions on this subject are so remarkable that the assertion of them would not and ought not to be credited, unless supported by a very considerable number of citations. There is unfortunately no lack of passages to be cited. Mr. Lewes is so ill-disposed to Things-in-themselves *et id genus omne*, that he coins a new word to stigmatise them, a term of contempt as he is careful to inform us,[2] namely, the *Metempirical;* and for the delusive science that has not shaken itself free from them, the term *Metempirics.* "Physics and Metaphysics deal with things and their relations, as these are known to us, and as they are believed to exist in our universe; Metempirics sweeps out of this region in search of the *otherness* of things: seeking to behold things, not as they are in our universe—not as they are to us—it substitutes for the ideal constructions of Science the ideal constructions of Imagination."[3]

---

[1] An extremely able criticism of Mr. Spencer's Theory of the Absolute, from the same reflective point of view as my own, is to be found in Mr. Alfred Barratt's *Physical Ethics*, 1869. App. I. p. 299.

[2] Problems of Life and Mind, Vol. I. p. 17.

[3] The same, Vol. I. p. 18. See also p. 27-28.

Book I.
Ch. III.

§ 3.
Scientist
theories of
them.

Well, this notwithstanding, we shall find Mr. Lewes holding, first, that the metempirical *exists*, though it is unknowable; secondly, that the metempirical does not exist at all; and thirdly, that the metempirical both exists and is partly knowable. I shall group the passages which I cite in support of these assertions under these three heads.

1. *Passages in which the existence of the "metempirical" is asserted, though unknowable.*

"Kant displayed great ingenuity in proving that the empirical and metempirical worlds (by him called the phenomenal and noumenal) having nothing in common, no conclusions formed respecting the one could have any validity when extended to the other. Why then did he continue to coquet with Metempirics, after having struck such blows at its foundation?"[1] To identify the metempirical with Kant's noumenal (supposing this to be really intended) would be to assert the existence of the metempirical, though unknowable, since that is the character of Kant's noumenal. The present passage is not decisive on the intention. We shall shortly come to one in which the identification is much more explicit.

At Vol. I. p. 28. the "Supra-sensible" is identified with the "otherness of things;" and it is said, that the knowledge of it, even if gained, would still be "relative, phenomenal." Again at Vol. I. p. 29 31. the metempirical is identified with the unknowable; and at p. 31. it is said, that conjectures in that region "may be approximately right, or absurdly wrong." I argue, then, that, if they may be approximately right, this involves the supposition of an object-

[1] Problems of Life and Mind, Vol. I. p. 19.

THINGS-IN-THEMSELVES AND PHENOMENA. 191

BOOK I.
CH. III.

§ 3.
Scientist
theories of
them.

matter for them existing. At Vol. I. p. 33. questions about the metempirical are said to be "*adjourned, not suppressed.*" At Vol. I. p. 252. we read, " I am far from implying that a Supra-sensible does not exist." And at p. 253, the field of speculation is divided "into the Sensible World, the Extra-sensible World, and the Supra-sensible World: a division corresponding with our previous distribution of positive, speculative, and metempirical."

At Vol. I. p. 182-3. we have a still more distinct assertion of the existence of the supra-sensible and unknowable. It occurs in the section headed *Reasoned Realism*, which is a discussion anticipatory of one more complete, probably that which appears in Prob. VI. in the second volume. "The Reasoned Realism of this work denies altogether the assumed distinction between noumenon and phenomenon—except as a convenient artifice of classification by which the *unknowable otherness of relations* is distinguished from the *knowable relations:* that is to say, noumena standing for things in their relations to other forms of Sentience (if there are such) than our own; and phenomena standing for things in any conceivable relations to Sentience like our own. Getting rid of the *Ding an sich*, or noumenon, as a phantasm that has no existence for us, consequently cannot come within our perceptions, nor within any theory of perception, and is therefore altogether banished from the sphere of knowledge, we are led through our psychological analysis back to the synthetic starting point—namely, that the external world exists, and *among* the modes of its existence is the one we perceive." [1]

[1] Problems of Life and Mind, Vol. I. p. 182-3.

Book I.
Ch. III.

§ 3.
Scientist
theories of
them.

Kant's position precisely. *limit-concept* included, as it appears in that Chapter, analysed above, on phenomena and noumena. I do not say that Mr. Lewes reaches his position by the same steps as Kant, but merely that his position is identical with Kant's when reached. The position is—that an Unknowable is known to exist, Kant's noumena in *negative* sense. But Mr. Lewes does not seem to be aware of the distinction between noumena in *negative* and noumena in *positive* sense; and when he argues against their knowability seems to have the latter in his mind. Mr. Lewes expressly repudiates Kant's doctrine of mental forms, in a passage of which the following is a part: " But as the sentient Organism develops, the external Order emerges; not because this Order is the creation of the Organism, *stamped upon* the chaos, but because this Order is *assimilated* by the Organism.—selected. according to its shaping reactions, from the larger order of the Real."[1] Observe, the *larger* order of the *Real;* just as above. " *among* the modes of its existence."

At the end of this section on Reasoned Realism, we come to a passage,[2] already cited in a former Chapter, in which the two senses of the word Object are noticed, being, in the first sense, the *other side* of the Subject, in the second, the *larger circle* which includes it.

Other passages which assert the existence of the Supra-sensible are: Vol. I. p. 370-1. where mystery is said to be not excluded from the universe but only from science: Vol. II. p. 1. " The Universe is mystic to man, and must ever remain so; for he cannot tran-

---

[1] Problems of Life and Mind, Vol. I. p. 184.

[2] The same, Vol. I. p. 189.

THINGS-IN-THEMSELVES AND PHENOMENA. 193

Book I.
Ch. III.

§ 3.
Scientist
theories of
them.

scend the limits of his Consciousness, his knowledge
being only knowledge of its changes;" Vol. II. p. 8.
"Unexplored remainders lie beyond every limit."
Again, Vol. II. p. 247. "whatever lies outside this
range" [*viz.* : sensible and extra-sensible experiences,
individual and social] "may belong to the Universe,
but not to our Cosmos, not to the *knowable* Universe,
and it is therefore rejected from Research." At Vol.
II. p. 364. again, there is a remarkable passage : " We
must not blow hot and cold with the Unknowable.
We must not pretend to any knowledge whatever of
it. If we are compelled to admit an existence which
is inaccessible" [well, what follows on this Kantian
admission ?] " we are not thereby compelled to doubt
the validity of our relative knowledge." I shall recur
to this passage again. But meanwhile, note the conception that the relativity of knowledge implies an
existence beyond the reach of that knowledge, a conception belonging to direct, not to reflective, consciousness.

2. *Passages which deny the existence of the "metempirical."*

The only passage in Vol. I. which I have been
able to find is at page 362. "Speculation craves for
a vision of the thing, or event, *in itself*—i.e., *unrelated:* in other words as it does not and cannot
exist." Does not and cannot *exist.*

In Vol. II. at page 20. " Philosophers" are said
to be repeatedly misled into the belief that their artifice has "its parallel in reality," and assign " a
reality to negative conceptions." At p. 21. a distinction is drawn between "the correlatives which
are logical and the correlatives which are real;" and

Book I.
Ch. III.
§ 3.
Scientist
theories of
them.

at page 22. it is said, that "between Non-Being, Pure Space, and the *Ding an sich*, there is no intelligible difference, except such as each borrows from its correlative."

At Vol. II. p. 26-27. "Nothing exists in itself and for itself; everything in others and for others: *ex-ist-ens*—a standing out relation. Hence the search after the *thing in itself* is chimerical: the thing being a group of relations, it *is* what these are."

But there is yet a more direct denial of the existence of the Thing-in-itself; it occurs in Prob. VI. Ch. II. a chapter referred to at the passage which has just been cited: "Its existence is not to be granted. It is a fiction and we know its genesis."[1] Nothing can be more distinct than this denial; and I will add, that if this second series of passages was all that Mr. Lewes had said on the subject, it would, though very incomplete, have given no cause for positive dissatisfaction.

3. *Passages in which the "metempirical" not only exists but is also partially knowable.*

At Vol. II. p. 12. "We postulate an indefinite Unknown." An unknown, not an unknowable; but then *indefinite*. At p. 124. "The reader sees that I am here speaking of Nature *not* as presented and re-presented in Sense and Thought, but as the pure Existence, the ultimate Reality, believed by all except idealists to *exist* independently, though only felt and known under subjective conditions: the postulated macrocosm which in us is a microcosm; the Universe as distinguished from our Cosmos ; or, to word it differently, the Sum of Things, as logi-

---

[1] Problems, &c. Vol. II. p. 439.

BOOK I.
CH. III.

§ 3.
Scientist
theories of
them.

cally distinguished from that portion which is comprised in our feelings." But how can this be *believed* to exist, unless it is in some way or other "comprised in our feelings"? What justifies us in "postulating" it? No answer from Mr. Lewes, nor is a satisfactory one possible on his method, unless the following should be thought so: "Our Cosmos is indeed the Cosmos of Feeling; but we postulate an universe of Being; and the warrant for this postulate is the experience of ever-fresh accessions from the Unknown to the Known."[1] But according to the passage just cited from page 124. it is "the pure Existence, the ultimate Reality;" "Nature, *not* as presented and represented in Sense and Thought" (that is, the Unknowable), which is postulated. It seems that the indefinite unknown and the unknowable must be much the same thing to Mr. Lewes. Accordingly we need not be surprised to read at page 252. "Much that is transcendental to-day may become empirical to-morrow; * * * the horizon of Experience is a movable and moving boundary. But so long as such transformation is not effected, whatever is metempirical is excluded from research." Until we know something definite of a thing, that thing is held to be not unknown but *unknowable*, for that is meant by "metempirical," as we have seen. Yet the admission that it *may* become known to us,—is this consistent with its *unknowability?*

At Vol. II. p. 292-3. there is an instructive passage: "When therefore it is argued that the creation of Something from Nothing or its reduction to Nothing is unthinkable, and is therefore peremptorily to be rejected, the argument seems to me

[1] Problems, &c. Vol. II. p. 235.

BOOK I.
CH. III.

§ 3.
Scientist
theories of
them.

defective. The process is thinkable but not ima-
ginable, conceivable but not provable." That is to
say, conceiving or thinking has a wider range than
perceiving or imagining. This is strange doctrine
from Mr. Lewes, who has argued that conception
depends upon and is symbolic of perception.[1] This
larger relation of concept to percept is the strong-
hold of Kant, of Hegel, and of Mr. Spencer. But
with so much variation in Mr. Lewes' conclusions, we
must not be surprised at some little variation in his
premisses.

Mr. Lewes' Prob. VI. is that which contains his
chief argumentation against Mr. Spencer. In the
first Chapter of that Problem he says: "The contention of those who declare the Absolute to be unknowable is, that beyond the sphere of knowable
phenomena there is an Existent, which partially *appears* in the phenomena, but *is* something wholly
removed from them, and in no way cognisable by us.
This may be so; but we can never know that it is so."[2]
Strange as it may seem, this contention, so precisely
like what Mr. Lewes has himself maintained in passages already cited, is what he now addresses himself
to dispute. Well, he does dispute it, and disputes it
successfully in my opinion. He shows what he asserts at page 425.—"that our knowledge of the Absolute, so far from being hopeless, is wide, varied, and
exact;" meaning that our *empirical* knowledge *is* a
knowledge of the Absolute. But the question of
course is,—What of the Supra-sensible part of the Absolute? Does Mr. Lewes say or imply, that we have
any knowledge of this? I will add two more passages

[1] Problems, &c. Vol. II. p. 16. et seqq. "*The twofold aspect.*"
[2] The same. Vol. II. p 130.

THINGS-IN-THEMSELVES AND PHENOMENA. **197**

Book I.
Ch. III.

§ 3.
Scientist
theories of
them.

to those already cited, in proof that he does. The first is from page 453. "I foresee an objection which some of my readers may raise, namely, Is not the Absolute the unknown quantity of which phenomena are the functions? It is thus conceivable. But observe, when y is said to be a function of x, and varies with it, we assume a knowledge of the variations of x, although ignorant of its numerical value. That is to say, unless x is akin to y in following the same numerical laws, we cannot operate on it through y." By analogy, then, the supposed unknowable part of the Absolute is akin to its knowable and known parts. There is a continuously varying connection between them. This is again stated in the concluding passage of the Problem, page 502-3. "Existence —the Absolute—is known to us in Feeling, which in its most abstract expression is Change, external and internal. * * * There is no real break in the continuity of Existence; all its modes are but differentiations."

Mr. Lewes has thus got himself and his readers into a very prettily complicated labyrinth, and without any suspicion that it is one. Let us see, if we can, by what steps he was probably led on into its mazes. The first error was, I imagine, the adoption of the usual Nominalism, common in English philosophy, as a sufficient basis; that is, he begins by considering what sorts of knowledge can *profitably* be entered on. The supra-sensible, his "metempirical," is clearly to him unprofitable and vain. But he does not dream as yet of denying its existence. He is bent solely on proving its unknowability. He therefore accepts

Book I.
Ch. III.

§ 3.
Scientist
theories of
them.

unhesitatingly the aid offered by identifying it with Kant's *noumenon*. But having admitted it to exist, he is logically compelled to assert something about it; it exists, and therefore cannot be *unrelated*, for to be unrelated is not to exist. It exists then, and in relation, is *partially* known to us, and continuous with the known. This contradicts his original assertion of its unknowability. At the same time, as identified with Kant's *noumenon*, it is for Mr. Lewes *unrelated*, a fiction, a member of a merely logical distinction of correlatives. Hence it exists and does not exist; it is unknowable and yet partially known. The identification of the supra-sensible with Kant's *noumenon* seems to me the thing which has wrought the mischief. Kant's remark about the *mundus sensibilis* and *intelligibilis* might have served as a warning against this; and so might his distinction between Noumena in a negative and Noumena in a positive sense.

In short, to identify the things of the Unseen World, Mr. Lewes' Supra-sensible and Metempirical, with Things-in-themselves involves contradictions. Whatever exists is phenomenal; and things-in-themselves are a fiction. But this is not to remove the natural limitations of human knowledge. Abolish the things-in-themselves, and what remains is a distinction *within* the phenomenal universe, falling *between* its seen and its unseen portions.

There is accordingly but one way out of that maze of contradictions; it is to assert both existence and partial knowability about everything, including even the "supra-sensible" and "metempirical," though we need not adopt these names as descriptions; and at the same time to draw a clear and firm

Book I.
Ch. III.

§ 3.
Scientist
theories of
them.

distinction between what is designated by those names, which is in truth the content of the unseen portion of the universe, the object-matter of the constructive branch of philosophy, on the one hand, and *Things-in-themselves*, as a pure delusion, on the other.

This is the position to which the method of philosophy, the method of reflection, leads. This method alone really abolishes the *Ding-an-sich*, at the same time abolishing the distinction between phenomena and noumena. That which *was* the *Ding-an-sich* becomes phenomenal, and the *Ding-an-sich* becomes a phrase without meaning, a fiction, a non-entity. That which was its entity, that which was denoted by it, is now brought within the phenomenal. The distinction itself has vanished. You cannot keep it, otherwise than as the record of a mistake, and at the same time deny the existence of *noumena*. There are, as Mr. Lewes well points out, merely logical distinctions; there are logical correlatives, the existence of which, except their momentary and untrue existence as fictions or attempts at thought, is not implied by drawing the distinction which sunders them; and *phenomenon* and *noumenon* are a pair of this sort. To draw a distinction at all involves assuming that the terms have a meaning, and the things an existence, for the purpose of argument. We shall see more about this process when we come to the question of the logic of possibility and apagogic proof. If either of the terms assumed for argument's sake is found, on examination, to have no meaning at all, to involve a self-contradiction, it is then emptied of all connotation, and the thing found not to exist. To empty it of all meaning is therefore to abolish and retract the dis-

Book I.
Ch. III.
———
§ 3.
Scientist
theories of
them.

tinction itself. It is a logical artifice which has served its purpose.

Another instance is the distinction between Being and Non-Being; it is the way in which we express that Being *cannot* be transcended. Nor can the use of such distinctions as these be set down to our conceiving what we cannot imagine. We can neither conceive consistently nor imagine consistently Non-Being; but we can *try* to do both, and mark our effort by *words*. Words are an instrument which go both beyond consistent thought and beyond consistent images; like certain symbols in algebra, those for imaginary quantities. Absolutists like Hegel may build these "attempts" into their system, on the ground that language is a derivative of thought. So it is, and of wrong thought as well as right. You cannot build Non-Being into your system without building self-contradictions into it. And why? Because you are taking the *pure* form or principle of thought, *contradiction*, taking it *as* pure form, and yet ascribing to it a *content* as such; thus beginning with a concrete contradiction, a thing which it is the sole use of the *pure* principle of contradiction to avoid.

Non-Being and *Noumenon* have, it is true, for the moment of distinction, a *real* existence; but it turns out that this existence is self-contradictory; a *real* contradiction,—an *untrue* existence; and the reality is the guarantee of the falsity. Mr. Lewes is therefore on the right track in the second series of passages, where he denies the existence of the Thing-in-itself. But then why does he identify it with the Supra-sensible?

Existence is the correlate of consciousness, both terms being taken in their largest and most general

Book I.
Ch. III.

§ 3.
Scientist
theories of
them.

sense. All definite knowledge is a part of this large consciousness, and its objects of this large existence. But this simple key to the puzzle of *noumena* is not grasped by Mr. Lewes. Indeed, it is the very purpose of his book to substitute for it the method of science in dealing with the problems of philosophy. It is, then, from no desire to convict a distinguished writer of inconsistency, that I have brought the details of it to light, but in order to show the necessity of the true philosophical method by contrasting its success with the failure of that method which is proposed as a substitute. And I will appeal to the candour of Mr. Lewes himself, whether it is not equitable, to say nothing of the necessity of the case, that one who has passed over into the metaphysical camp only to surrender it to the scientists, his former comrades, should furnish if possible from his own writings the proof, that the contemplated surrender has not hitherto at least been effected. The question of Noumena, or Things-in-themselves, is one of the most fundamental in philosophy, as well as one which most constantly recurs. No one who has not the key to its solution can pretend to have the key to philosophy generally. This question once solved, but not before then, we may have a legitimate confidence in approaching the remainder.

There is no doubt a very real and tangible difference between Mr. Lewes' doctrine and Mr. Spencer's. To the latter, the unknowable real is more real than phenomena; to the former it is not so. To the latter, phenomena depend for their existence upon the unknowable real, to the former not so. And Mr. Lewes at least holds that while his unknowable real does not in any way affect the validity of

BOOK I.
CH. III.
_____
§ 3.
Scientists
theory of
them.

actually acquired knowledge, Mr. Spencer's does so; also, while Mr. Spencer conceives his unknowable as the Object of Religion, that of Mr. Lewes is not capable of affording religion a basis.[1]

It seems, however, to matter comparatively little whether we regard the "indefinite unknown" as lying *beyond* the sphere of our actually known phenomenal world, or as lying *behind* it, in the character of a substrate or cause. I mean, it matters little in respect of the worth and validity of our actually acquired knowledge, our knowledge of the *mundus sensibilis et intelligibilis*. For consider how very minute a portion of the indefinite unknown this *mundus sensibilis et intelligibilis* must be; and how much the *character* of its laws must depend upon their connection with the laws of the larger universe, even though the content of those known laws may remain true for ever, so far as it goes. "The validity of our relative knowledge" which Mr. Lewes is so anxious to save, in a cited passage[2] to which I said I should recur, is not saved more effectually on his theory than on Mr. Spencer's, nor, I will add, than on the philosophical theory which includes the "suprasensible." All three theories really include the things which two of them profess to exclude by calling them the *Unknowable* and the *Metempirical*. And for all three theories alike these things are the vaster, the preponderating, portion of the universe of existence, infinite where that is infinite, unknown where that is unknown, but continuous with the limited and finite portions of existence, and with that which constitutes our seen world.

[1] See Problems, &c. Vol. II. p. 451-453.

[2] The same, p. 364. cited at p. 193.

Book I.
Ch. III.

§ 3.
Scientist
theories of
them.

If it is remarkable that Mr. Lewes should suppose himself to differ so profoundly from Mr. Spencer, it is still more remarkable that he should suppose himself to differ at least as profoundly from Kant. " All modern Metempirics is either Kantian, or founded upon Kantian principles," he writes,[1] and proceeds to criticise "Kant's fundamental positions." The differences are no doubt great; but there is this overwhelming similarity, that both begin by distinguishing Phenomena and Noumena, then assert the *existence and unknowability* of Noumena, and end by bringing them back into knowledge again. Mr. Lewes has added the term *Metempirical.*

Kant's originating error with regard to Noumena was not, as Mr. Lewes seems to think, in phenomenalising them after introduction, but in introducing them at all. Once admitted, there is no help for it but to furnish them with predicates of some sort. This originating error Mr. Lewes shares, and consequently in spite of his disclaimers of the possiblity of knowing anything about them, disclaimers not a whit stronger than Kant's in the Critic of Pure Reason, he also shares the error (if it be one) of phenomenalising them after their introduction. Mr. Lewes, however, has not Kant's excuse for their introduction. They flowed almost inevitably from Kant's general analysis, or diagnosis, of consciousness; he has the great merit of giving a theory of them based on that analysis, and thus bringing their nature out into the light, and rendering their existence or non-existence a verifiable assertion. And I have already pointed out that their non-verification entailed, by a *reductio ad impossibile,* the ruin of Kant's analysis of consciousness, from

[1] Problems, &c. Vol. I. p. 437.

Book 1.
Ch. III.

§ 3.
Scientist
theories of
them.

which they flowed by a necessary consequence. But
Mr. Lewes denies Kant's analysis of consciousness,
and yet introduces Noumena,—*gratuitously* encumbers
himself with them. Why, if I may repeat the remark, did he not take the warning which Kant gives
in the Chapter of the Critic analysed above, that the
distinction between phenomena and noumena has
nothing to do with that between the *mundus sensibilis*
and *mundus intelligibilis*, which latter distinction does
not belong to the system of the Critic ? It is to these
that Mr. Lewes' Sensibles and Extra-sensibles correspond; and in this series his Supra-sensibles make
a third member. Mr. Lewes, then, should either have
said nothing about Noumena, Things-in-themselves,
at all, or else, if he was bent on solving that problem,
should have approached it by the philosophical method.

But it will perhaps be replied to this criticism,
that, quite independently of Kant's theory or any
theory at all, there does arise in the understanding
the fallacious notion of Things-in-themselves, and
that this notion is most difficult to explain and eradi-
cate. The notion of Things-in-themselves have then,
as a matter of fact, not that origin which Kant's
theory assigned to them, but some other probably
more simple origin. To combat the notion, therefore,
it is requisite to show what its origin really is, and
why it is fallacious. This is perfectly true, and for
this very reason it was that the present Chapter
began with a diagnosis of the origin of the notion;
and thus too Mr. Lewes, to whom Kant's account of
its genesis is equally baseless, gives an account of the
way in which he supposes it really to arise in the
mind. It comes, he says, from the tendency to ima-

BOOK I.
CH. III.

§ 3.
Scientist
theories of
them.

gine single things wherever we have single names, and in the case of concrete things to personify or sub-stantialise the abstract thing apart from its qualities and properties. " We abstract this *Also*, personify it, assign it an imaginary substance, and assume that the Possibility is a Reality apart from all conditions."[1] If then we can show that this Thing-in-itself, so formed, is a non-entity, a fiction; which we can do by analysing the several cases in which it is supposed to exist, and by showing that there is nothing in the thing analysed that is not fully accounted for by the members of the analysis when performed; that is, by showing that we have exhaustively analysed the thing without residuum; then we shall have destroyed the fiction beyond hope of recovery, and have solved the so-called problem of Noumena.

This is a way which Mr. Lewes has actually taken in two instances at least, and with very marked suc-cess. The two instances, as I have before remarked, are Matter and Force, in his two Problems of Vol. II. entitled *Matter and Force*, and *Force and Cause*. This method of beating the Thing-in-itself in detail is effective where it is possible. But it is not always possible. In analysing matter and force we have all the strength of modern science to aid the investiga-tion. When it is *entia rationis*, "extra-sensibles," that have to be analysed, it may not be so easy to apply this method. But though we have the aid of science, still even here the method is not that of science but of philosophy, the method of reflection. Follow Mr. Lewes at each step of his brilliant analysis in these two Problems, and you will find him con-tinually asking what is the *meaning* of this term and

_____
[1] Problems, &c. Vol. II. p. 440. And see the whole Chapter.

Book I.
Ch. III.

§ 3.
Scientist
theories of
them.

that, what do we *feel* when we have this or that experience, what are the distinguishable *elements* of this or that perception.   He is not testing facts by experiments in the laboratory, but testing concepts by comparing them with percepts, or, as I should say, testing *second intentions* by comparing them with *first intentions*.

But effective as this mode of operation is, and has been shown to be by Mr. Lewes' able and successful application of it, yet it is not a mode of operation which can solve the *general question* of Noumena. You may prove that there is nothing answering to this or that particular Noumenon, the noumenon Matter, or the noumenon Force; but you will be far from having shown that there is nothing answering to the combination of Object and Subject.   Mr. Lewes has attempted this, but as it seems to me, and as I have tried to show, without success.   His Object-Subject gives a "Monism;" and then even this is only a part of the larger total of Existence.[1]   To solve the general question, and explode Noumena altogether, the *recognition* of the philosophical method is requisite.   And it is so for this reason, that the root of the evil lies much deeper than Mr. Lewes supposes.   It is no mere tendency to personify groups of qualities expressed by single names; no tendency to substantialise abstractions by itself alone, though this is also a powerful contribuent.   This is a kind of error which can be rectified when it is once clearly exposed.   It will not account for the tenacity with which carefully elaborated theories, like that of Mr. Spencer for instance, are held and insisted on.   No; the tendency to believe in Things-in-themselves must

[1] Problems, &c. Vol. I. p. 122.

BOOK I.
CH. III.
———
§ 3.
Scientist
theories of
them.

have a far deeper seat than this; a seat no less deep than the tendency to be satisfied with the method of direct instead of reflective consciousness. And if this is the origin of the notion, it is clear that no beating it in detail can be a sufficient defeat. For you will have combated an effect, a symptom, of the notion, and not the notion itself, which is not many but one. The method of reflection alone, which is the method of philosophy, combats the notion itself, by showing the contradiction involved in it, and pinning it down to its definition—*an unknowable existent.* But from the extreme generality of this process, supplementary disproof in detail of particular cases of noumena is practically indispensable. Such particular and *a posteriori* disproof as Mr. Lewes has given, in the cases of matter and force, is valid on any theory of the origin of the notion of Things-in-themselves. It is not to be mixed up with the general question, to which the origin of the notion is a vital point. If the account of Things-in-themselves which I have sought to establish be the true one, the belief in them will not disappear from speculation except where and so far as the method of direct perception is combined with, and in philosophical questions subordinated to, the method of reflection.

§ 4. But it is time to clothe these dry bones of analysis with the flesh and blood of human interest, and to consider what is, in outcome, the drift and scope of the whole question as to Things-in-themselves. It is this. The question really means, What is the ultimate and most secret *source of being;* what is the power in virtue of which existents severally, and even the seen world as a separate whole, exist? The attempt to answer this question by subordinating it to the distinction between Things-in-themselves and Phenomena, and then asking what are Things-in-themselves, was radically vicious, because the distinction itself was false. The attempt to answer the question suffered from an incurable logical vice, but the question itself was legitimate. The source of being, the cause or power by which existents (severally taken) exist, are all terms which have a phenomenal meaning, though it may not be one which it lies within the grasp of human faculties to determine more precisely than is already done by the questions themselves. Suppose an answer to have been obtained;—the source, the power, now determined, will not be a Thing-in-itself, but a Phenomenon, an existent with a subjective aspect. By obtaining the answer and determining the source in question, the field of direct consciousness will have been enlarged by ground which was previously covered only by reflective; a possibility of thought will have become an actuality of knowledge. By denying the *existence* of the Unknowable, of Things-in-themselves, we are not restricting the field of existence but redistributing it. Reflective consciousness includes what *was* called unknowable existence.

Many people argue for Noumena thus: *Phenomena* cannot exist unless *Noumena* do, for they are relative terms; phenomena therefore *suppose* noumena. My view on the other hand is, that what *were* thought to be noumena are by reflective analysis shown to be really phenomena. True, the terms are relatives; but they indicate now, after analysis, the history of the single class, phenomena, which has the other class, noumena, subsumed under it. The terms must still be used as relatives, in order to show that in the " existence," which now contains both, phenomena are not classed as a case of noumena, but noumena as a case of phenomena. Take an illustration. Mountains and valleys are relative terms. Suppose a levelling of the country. The distinction will equally disappear, whether the valleys are brought up to the mountains, or the mountains down to the valleys. In the first case we shall have a table land, in the second a low lying plain. The sea level is the common standard in respect of which the plain and the table land are now compared. Such a common standard exists also in the case of phenomena and noumena; it is the principle of reflection. The term *phenomena* has a meaning with respect to this principle, quite irrespective and independent of the existence and meaning of the term *noumena*. It means existence with a subjective aspect, things seen or known. *Phenomena*, then, do not in the least depend for their existence on *noumena*, notwithstanding that the terms are relatives. They are the table land to which the valleys have been brought up, mountains and valleys, *as they were*, being alike abolished.

It was Kant who, by following out his *transcendental* point of view, made the transition to the new conception possible. He gave a theory of the absolutist or realist notion which lurked in the direct method of regarding existence. He forced this lurking absolutism, this lurking realism, to appear in the shape of the Thing-in-itself. The realism which had been held in solution was suddenly precipitated, and what was it? A contradiction in terms.

The philosophical world was in an uproar. How was it to deal with its troublesome present, the Thing-in-itself? For Kant did not know. He had raised the ghost but he could not lay it, because he was not closely enough acquainted with the nature of the summoning spell which he had himself used.

The years which followed the publication of the *Critic of Pure Reason*, from 1781 to the end of the century, are the period in the history of philosophy which is the most practically interesting to us at the present day.[1] It witnessed the striking out of the three directions in which progress from Kant in respect of Things-in-themselves was possible. There were but three; for Herbart's Realism, though founded on Kant's conception of existence, was really a retrogression to pre-Kantian ideas; because the conception of existence as mere *positing* (*Setzung*) misses its essential element, the *reflective* character of the positing, and is nothing but an assertion of that very contradictory notion which Kant himself had rendered untenable.

Of the three courses which were really open, the first was to accept the self-contradictory Things-in-themselves, and boldly affirm them as the objects of Belief though not of Knowledge. This was Jacobi's line. The second was to take Kant's diagnosis of consciousness as it stood and carry it out into a constructive philosophy. This was Fichte's course, in which he was followed both by Hegel and by Schelling, each with characteristic differences. This also was the course taken by Schopenhauer, basing him-

---

[1] See Herr Kuno Fischer's admirable account of the development between Kant and Fichte, in his *Geschichte der neuern Philosophie*, Vol. V. pp. 1. to 212.

Book I.
Ch. III.
§ 4.
Scope of the
whole question.

self, however, on a different principle from theirs, namely, on the *active* element in consciousness, the Will, as opposed to the perceptive or intelligent element. The third course was to carry on the critical and analytical methods of Kant to still further results; and this was done by Salomon Maimon, who for this reason is Kant's true successor in the field of Metaphysic, the analytical branch of philosophy. Maimon is the first and greatest of modern metaphysicians. To him we owe the preservation of the continuity of metaphysical tradition, its continuity with Kant, the last of the great founders.

The Thing-in-itself had to be dealt with; that self-contradictory source of being had to be accommodated. Now the world was naturally in a hurry,—as it always is. It wanted some tenable theory for immediate use. It could not afford to have its intellectual basis of life in a state of dissolution. A *source of being* it must and would have, honestly if it could, but a source of being. Who prevented it? Only Kant. Well, then, either through Kant or in spite of him a source of being must be obtained. In spite of Kant, said Jacobi, that is, as an *object*, independent of us. Through Kant, said Fichte, that is, as a *subject*, as our Self. Fichte's "absolute ego" is an "ego in itself," an *Ich an sich*. He identified subject with object, thus satisfying the claims of reflective consciousness; and he identified this identity with the source of being, that is, with the Thing-in-itself. The *ego*, in short, was all that the Thing-in-itself professed to be, namely, a source of being,—and at the same time was purged of its self-contradiction, inasmuch as it was also the source of knowledge.

It was a masterpiece of inventive genius, this

clumping together the two aspects, subjective and
objective, and then clumping (together with them)
the analysis and the source of existence; making the
two aspects serve as an *analysis* in their character of
opposites, and as a *source* in their character of iden-
ticals. It is a repetition of the clumping trick of
Leibniz and of Spinoza, only on a more complicated
stage. As Leibniz clumped force and consciousness
in his Monads, and Spinoza extension and thought
in his Substance, so Fichte activity and self-consciousness in his Ego. And it is essentially the same
conception which lies at the root of Hegel's theory;
only that his "Absolute Mind" is generalised, instead
of remaining, as in Fichte, a plurality of individuals.
As Hegel says in the Preface to his *Phänomenologie*,
" Everything depends,—according to my view, which
must justify itself only by the complete exposition of
the system itself,—everything depends on appre-
hending and expressing the True not [only] as *Substance* but equally as much (*ebenso sehr*) as *Subject*."[1]
In both Fichte and Hegel the " Absolute" is a process working by the logical law of contradiction; objective and subjective at once, existence and the
source of existence at once. Fichte's theory is Leibniz', Hegel's Spinoza's, in motion.

It was a masterpiece of inventive genius. But it
was not analysis, and it was not " *Criticismus.*" In
contrast with this ambitious new Ontology, and in
contrast also with the old ontology of Jacobi, Maimon's humbler and more patient analysis might well
appear, what he himself called it, a " critical scepticism." It was scepticism in respect of that Thing-
in-itself which Kant again brought in by means of

[1] *Werke*, Vol. II. p. 14.

the Practical Reason; and it was scepticism in respect of that Thing-in-itself which Reinhold found in Kant and made the basis of his *Elementarphilosophie.* It was scepticism also in respect of the old absolutism, the pre-Kantian realism, which Maimon finally evaporated, by driving home the proof of the non-existence of the Kantian Thing-in-itself, the last form in which that old absolutism had taken refuge. Scepticism in respect of nonentities, it is the very reverse of scepticism in any serious meaning of the term. Maimon is sceptical, much as a Wiclif or a Luther is irreligious. He leaves unexplored but not self-contradictory the vast fields of possible existence and knowledge, of which we may well suppose our actually known world to be a portion, and on which, in that case, it must be conceived to depend as part of its conditions of existence.

Our actual phenomenal world is a part of a larger, but still phenomenal, world which we must conceive as possible, possible in our view, but actual to other modes of consciousness than ours. In other words, what *was* the Thing-in-itself has been phenomenalised and relegated to this possible world, has become an object of the constructive branch of philosophy. The *notion* of Thing-in-itself being abolished, the things to which it was applied are included under another notion, that of the Unseen World. The things are real phenomena; therefore, to abolish the one notion is to call the other into existence.

Let us note how this change has been effected. It is by a change from that feature of Kantianism which regarded objectivity as given by the Categories, to one which regards objectivity and subjectivity alike as determinations introduced into primary consciousness by reflection. Objectivity is given not by thought but by reflection. Therefore instead of the Thing-in-itself which is an "empty object," a contradiction in terms. we have a phenomenon which is beyond the range of direct, but within the range of reflective, consciousness. It is quite true that the same mode of consciousness which gives objectivity is also the condition of the notion of a source of being; but this mode of consciousness is, not thought as opposed to perception, but reflection as opposed to direct consciousness. Reflective perception gives objectivity; and reflective conception supervening upon it gives rise to the notion of a source of being. a question as to the source of the objective and subjective percepts.

It is evident that questions like that which has now been handled can be answered only on "critical" principles, by what Kant called the critical and transcendental, but what I, not holding his theories, prefer to call the analytical and metaphysical, method. The method and furniture of consciousness, its structure, content, and mode of operation,—these are the fundamental objects of philosophical enquiry; and to these I propose to proceed in future Chapters. It is to analysis of this kind that we must look for an explanation not only of what may prove to be true, but also of theories and imaginations which have been supposed to be true. concerning the filling up of that region of philosophy which constitutes its constructive branch: an explanation of the various shapes into which the problems of how the world came to exist. and the like, have been thrown by the human mind. From human nature, in its experiences inward and outward, and from no other source. flow all possible theories of the supernatural. All the fancies that have ever obtained credence among men are expliplicable, if at all, by this means. They have been pictures drawn upon the dark background of the Unknown; and their incredibility, where they are incredible, does not arise from their background being unknowable, (which it is not), but from the fact, of opposite significance, that we are gradually coming to know more and more of the conditions requisite to their formation, in coming to know more and more of the nature and functions of consciousness.

BOOK II.

ANALYSIS OF ELEMENTS.

CHAPTER IV.

PRESENTATION AND REPRESENTATION.

§ 1. Philosophy cannot be defined, or have its field marked out for it, by the problems which it can solve or reasonably attempt. Like every other distinct branch of knowledge, it is defined only by its methods and its objects. The questions or problems which it may be enabled to solve concerning those objects, and by following those methods, are another thing. But when we have once got our definition of philosophy by object and method, which has been sufficiently done in Chapters I. and II., then we are in a position to see what are the main problems, the solution of which it may reasonably propose to itself. And the enumeration of these main problems will serve to give a greater definiteness and unity of purpose to a philosophical work.

The main problems of the analytical branch of philosophy, that is, of metaphysic, are I conceive two. The first is that of Things-in-themselves. The solution of this has already been given, in the foregoing Chapter, a negative solution, dissolving the illusion, and showing that there is no Thing-in-itself, because there is no existence beyond consciousness. It is a solution which depends purely on the phenomenon of Reflection. Without going farther than that phenomenon, we are enabled to solve this first problem of metaphysic.

The second problem is identical in purpose with that which Kant proposed to himself, and which he formulated by the question, " How are synthetic *a priori* judgments possible?" It is by no means necessary that we should formulate the problem in the same terms as Kant did; but it is indispensable that we should by some means or other envisage the question as definitely as he did. It is the question in which the chief interest of philosophical speculation lies; we may have a genuine philosophy which fails to give the true solution, but such a philosophy, though genuine, will be barren of its richest fruit. Not to set this problem definitely before us would be to turn philosophy aside to false or fruitless issues.

The problem may be designated as that concerning the strict and inviolable necessity of the Law of Uniformity of Nature; of the principle of Ratio Sufficiens; or of the Law of Causality; to find if possible a metaphysical basis for the sciences, what Kant called a synthetic *a priori* law for nature, one which does not depend, in the last resort, upon the order of experience in phenomena, but which, having another source independent of that order, goes beyond the certainty derived from experience, and procures a certainty that the uniformity of nature not only never has been known to be violated, but that it never could have been and never can be violated.

Of course, if we fail in demonstrating this inviolable metaphysical certainty of the Uniformity of Nature, we do not therefore renounce the lesser empirical certainty of it, which comes from experience, and with which men of science are, for the purposes of science, content. That experiential certainty stands firm on its own grounds. But what philosophical enquiry aims at is, to discover a proof, by subjective analysis, of a greater certainty in the law, of an inviolable uniformity in nature, of what may properly be called an *absolute* uniformity, if only the word *absolute* is used as opposed to *incomplete* or *partial*, and not as opposed to *relative* or *phenomenal*.

I shall offer a positive solution of this problem in a future Chapter. It cannot come now; for it depends not on the phenomenon of reflection only (like that of Things-in-themselves), but upon the formal elements in consciousness, and especially upon the Postulates of Logic, the principles of analytical thought, in conjunction with the phenomenon of reflection. When these have been examined, I shall attempt to show that, taken together with the phenomenon of reflection, they give us proof of the absolute certainty and inviolability of the uniformity of nature. In other words, I shall demonstrate the equivalent and more than the equivalent of Kant's so-called synthetic principle of causality, without recourse to any synthetic *a priori* forms of thought. I say I shall demonstrate this; but neither empirical thought, nor the argument against what is commonly called the miraculous, will take much by the demonstration. Only one proof more will be added of the thorough distinctness, although in close interdependence, of the two domains of science and philosophy.

These two problems are the main questions of metaphysic. A third problem, belonging to the constructive branch of philosophy, may be mentioned here. This problem concerns the nature of the connection between the seen and the unseen worlds; and the final Chapter will be devoted to its consideration. The object-matter of my constructive branch of philosophy corresponds generally to Kant's *übersinnliches*, the problems of which are according to him three,— God, Freedom, and Immortality. I do not take these precise problems, nor in precisely the same way. Still the problem of the constructive branch, being taken as the connection between the seen and the unseen worlds, may be said to constitute the great interest and final purpose of metaphysic itself, not that metaphysic treats the problem, but that it furnishes the only means by which it can be treated; metaphysical analysis supplies the materials for constructive synthesis. It is fully in accordance with this, when Kant gives two definitions of Metaphysic, the first drawn from its final purpose: "The science which carries us from the cognition of the sensible to that of the super-sensible, by Reason;" and the second a school-definition, as he calls it, drawn from what is, not aimed at by, but actually done in, metaphysic: "The system of all principles of pure theoretical reason-cognition by means of concepts; or more briefly, —the system of pure theoretical philosophy."[1]

§ 2. I have thought it best to premise the foregoing remarks, because we are now entering on another division of the subject, and they will serve to throw light back on the path we have traversed, and forwards on the path to come. We are now coming

---

[1] Fortschritte der Metaphysik seit Leibnitz. Kant's Werke, Vol. I. pp. 488. 490. Ros. u. Sch. ed.

Book II.
Ch. IV.

§ 2.
Alignment
with Hume,
Kant, and
Maimon.

within hail of the knottiest portion of the work, that where the reader who has courage to enter on it with me must make up his mind to bear the real burden and heat of the day. But I have still many remarks to make, before I actually enter on that knotty bit of analysis, which is the real subject of the Chapter.

The analysis of primary, reflective, and direct, consciousness, in Chapter II., was the justification of the object and method of philosophy, as set forth in Chapter I. Three *modes* of consciousness were distinguished one from another; and the special object-matter of one of them, the reflective mode, was shown to be the special object-matter of philosophy; and this special object-matter of philosophy was shown to consist of two opposite, co-extensive, and inseparable aspects, *objective* and *subjective*, or, when named each for itself, *existence* and *consciousness*. And the reflective mode of consciousness, or briefly Reflection, is that which not only draws the distinction between the three modes, itself and the two others, but also distinguishes the two aspects in its own special object-matter. Reflection is the mode of consciousness employed both in the foregoing analysis, and in that which is now to follow.

But in what is now to follow, reflection will devote itself to the analysis of its *special* object-matter, that is to say, to the analysis of the subjective aspect of existence, to the analysis of consciousness, not of self-consciousness. It will abstract to a great extent from the distinction of the three modes of consciousness, as having been already treated, and as containing its own warrant of legitimacy; and will analyse the different processes of consciousness and some also of its different states, or kinds of content, from its own

Book II.
Ch. IV.

§ 2.
Alignment
with Hume,
Kant, and
Maimon.

reflective point of view. It will take those states and processes as they are *in consciousness* alone; in other words, they will be states and processes belonging to the primary and not to the direct mode. For the direct states and processes are nothing more than the primary, with the added hypothesis of a *nexus* or *substantia* behind them which makes them separate objects. And any and every hypothesis of this nature it is the very purpose of philosophy to exclude at first from its object-matter of investigation, in order that the origin and nature of the *nexus* or *substantia* may be seen, what it consists of, and whence it comes into the phenomena. The states and processes of primary consciousness are, then, the object-matter of the following analysis.

Observe the significance of this method, and this choice of object. My theory, like Kant's, lays Apperception, *Anglicè* Reflection, at the basis of philosophy; but it is without the causal efficiency which it had in Kant. Its metaphysic is without the notions of substance and cause, as notions to begin with. The genesis of those notions can be traced, and themselves shown to have a phenomenal, not a noumenal, an originated not an originating, character. But with Kant the Transcendental Apperception was originative; it was cause and consciousness in one. In the first edition of the Critic of Pure Reason this is very clear. The "productive imagination" works by synthesis, is a synthetic faculty. It depends on the transcendental unity of apperception working in the twelve modes or forms of unity, called the Categories. The transcendental apperception, which unified everything in consciousness, was a creative source, *quoad nos*; a fountain that "flung up momently the

Book II.
Ch. IV.
———
§ 2.
Alignment
with Hume,
Kant, and
Maimon.

sacred river" of consciousness, forming into a stream the scattered waters of sensation inner and outer. Kant's theory, then, clumps into one thing the two separate moments of consciousness and cause. And this is his Transcendental Apperception.

But it will be asked, fairly and indeed necessarily asked,—Where do you look for the cause, the substance, the agent, the conscious thing, (call it what you will), of consciousness? If you refuse to put together cause and consciousness into one thing, you can have no conscious Soul or Mind, as well as no conscious transcendental apperception or Ego. Or, in order to place the causal nexus somewhere, do you call the Soul, or the Ego, *a series of conscious states becoming conscious of itself as a series?* For, if you do, you will only be again attributing causality to consciousness, in the words "*becoming* conscious," and it will be just the same essentially, as if you fairly adopted the expression Soul or Ego or Transcendental Apperception.

I fully admit the necessity of the question, and the justice of the last remark. And my reply is this. I put the enquiry into cause, agent, source, force, or however may be expressed the notion of *what makes*, into a separate, and a subordinate, department of the enquiry. I place first subjective analysis, an enquiry into the *nature*, the τί ἐστιν, of things; and secondly and subordinately I place the enquiry into the *genesis* and the *history*, the πῶς παραγίνεται, of things. The first enquiry is a branch of philosophy; the second and subordinate one is a branch of science; the first is, in the case of consciousness, metaphysic, the second psychology.

This premised, (and the distinction between

Book II.
Ch. IV.

§ 2.
Alignment
with Hume,
Kant, and
Maimon.

Nature and History is one of the most fundamental in my whole theory), I proceed with my answer. The nominal definition I would give of the soul or mind is—*a series of conscious states among which is the state of self-consciousness.* And the agent or substance which becomes conscious, or in which resides the force of becoming so, or which *has* the states of consciousness, is not the series or any one or more of the states which compose it, but (in man) the brain or nerve substance. When we draw the above necessary distinction between nature and history, then the question so often put—Materialist or Idealist? is to be answered, in the first place, by the farther question, —Do you mean in philosophy or in psychology? For the two domains are essentially different; and those who answer this question with me will probably reply also with me to the first question :—Idealist (or rather *Reflectionist*) in philosophy ; Materialist in psychology, and indeed in all the sciences. The causes and the genesis of this and that individual conscious being, as well as of each and all the states and processes of his consciousness, depend upon matter in motion. And if you tell me that matter in motion is nothing but sensations in coexistence and sequence,—I reply, that this is an analysis of the *nature* of matter, not an account of its genesis or history. The first *cause* that we can discover, anywhere, is matter in motion; and, that we can analyse this cause subjectively, only shows the truth of my assertion that the domain of genesis, of history, of science, is subordinate to the larger domain of nature and philosophy. I do not profess to assign the prior condition, the *substantia* or cause or agent, of consciousness at large. I exclude that question from

Book II.
Ch. IV.

§ 2.
Alignment
with Hume,
Kant, and
Maimon.

metaphysic. And I say that, if a prior condition of that combination of states of consciousness which we call *matter* could be assigned, (which smaller question is not necessarily unanswerable), it would be by an insight into the Unseen World, by a theorem belonging to the Constructive Branch of philosophy. Materialism, then, which is worthless as philosophy, inasmuch as it gives no account of what matter and motion are, or in what the efficiency of physical causation consists, is the only sure standing ground in science, where the problem is, assuming these phenomena as given, to measure, weigh, and predict, their sequences and coexistences.[1]

Now, to take consciousness and its phenomena to examine, as if they were objects of direct, and not, as they are, of primary and reflective consciousness, is to treat them as objects of science and not of philosophy, is to clump together in them causality and consciousness, is to *assume* that they have force or causal efficiency in them. This would lead, and has led over and over again, to an *a priori* psychology. And with an *a priori* psychology (to say nothing of an *a priori* philosophy) metaphysic has henceforth nothing to do. Fortunately we possess a genuine *a posteriori* experimental psychology, a true science, which is daily yielding results of the highest value to many able and distinguished investigators. Fortunately for the world, and fortunately also for metaphysic; for metaphysic will derive from that psychology an independent support and verification.

So much as to the absence of causal efficiency from my metaphysical theory. Another point which

[1] See the admirable remarks in Lange, Geschichte des Materialismus, Vol. II. p. 7, and frequently, 2nd edit. 1875.

Book II.
Ch. IV.
§ 2.
Alignment
with Hume,
Kant, and
Maimon.

may throw valuable light on my method is the following. I lay reflection at the basis, as Kant laid apperception, but with me it does not work through any *a priori* forms of unity. There are, in my theory, inseparable formal elements, as well as inseparable material elements in consciousness; time and space, as well as feelings. But these formal elements are not attributed to the mind or to the ego, or to "the within" in contradistinction from "the without," in any way. They and the material elements alike come forward as inseparable elements of analysis in the states of primary consciousness. They are found there by reflection. Their being found there, and not put there, by reflection is what constitutes reflection derivative and not primary.

But with Kant, not only is apperception a source of causal efficiency (as we have seen), but it works in and through certain *a priori* forms of its own, prior to the "matter" which it receives "from without." Kant thus separated the inseparable elements. According to him there are sensible impressions coming from without, and these, when perceived, are perceived always in the *a priori* forms of intuition, springing from within, Time and Space. Observe the theory. We always *perceive* the sensation and the form together. And yet we are to *assume* that they are originally separate, having separate sources.

But even then, this union of sensation and form is not cognition. By itself there is no "unity" in it, "Unity" is first introduced into intuitions by apperception working through the Categories, which are *a priori* forms of thought. There is an act of synthesis, originating in the transcendental apperception, as the condition of all cognition. What I mean to

BOOK II.
CH. IV.
———
§ 2.
Alignment
with Hume,
Kant, and
Maimon.

insist on is this, that in Kant there is not only a causal efficiency ascribed to the apperception, but also an elaborate machinery of transcendental forms,—namely, forms of intuition, forms of thought, and principles of judgment,—in and by which this causal efficiency operates; and moreover that in one important respect this machinery of forms is *separate* from the matter with which it is inseparably combined, so far as consciousness goes; I mean in respect of its belonging to and springing from that which is causal in the apperception, while the "matter" springs from some source or other not included in that causal agency.

To all such *a priori* theorising as this my theory bids a long farewell. Reflection is with me, as it was with Kant, the basis of the whole of philosophy, but—and here is the important difference—*not* as the *causa existendi* of consciousness, *not* (which soon developed itself out of Kant) as the *causa existendi* of existence, but as a particular kind of *causa cognoscendi*, namely, as the perception of the τί ἐστι, the analysis of the *essentia*, of consciousness and its states. Reflection is reëxamination of the states of consciousness from which it is derived, of that series of states of which it is a prolongation. This is the significance of its relation as a derivative from primary consciousness.

The change which I make is simply this,—I explain the distinction between Object and Subject by referring it to the moment of reflection instead of to the Subject of reflection, as Kant did, thereby explaining a thing by itself. Kant's philosophical Copernicanism,[1] to adopt his own comparison, led him

———
[1] Preface to the Critic of Pure Reason, 2nd edit.

BOOK II.
CH. IV.
—
§ 2.
Alignment
with Hume,
Kant, and
Maimon.

straight into this inconsequence. Cognition that has
a priori forms is already a cognition-faculty, a Subject, and an Existent. It supposes the distinction
between Object and Subject already drawn, and instead of coinciding with that distinction, as the moment of reflection does, which is the perception of it,
it coincides with one or other of the two *membra
distinctionis.* Nor is this all; it coincides now with
one, now with the other; for, as an *existent,* cognition
is an Object, as faculty *of cognition* it is a Subject.
Kant's Copernicanism first made consciousness, instead of objects, causal ; and then made it an object,
because causal.

Existence is thus thought, according to Kant, by
virtue of an *a priori* form in the cognition-faculty *of
an existent;* an explanation of existence which assumes
the very notion to be explained. Reflection, on the
other hand, gives some further account of the notion
itself, though throwing no light whatever on the
question how the thing, existence itself, arises. The
genesis of existence at large is a contradiction in
terms.

Reflection, then, can only be the central point of
philosophy in the character of a moment or special
state of consciousness, as identified with the distinction between Subject and Object, and not with the
supposed Subject which draws that distinction. Otherwise you merely assume existence to account for
existence, and are then forced for mere consistency's
sake to theories like Fichte's, the *sich-selbst-Setzung*
of an Absolute Ego, simply in order that there may
be some difference between what you account for and
what you rely upon to account for it.

The notion of Existence being Positing, *Setzung,*

Book II.
Ch. IV.

§ 2.
Alignment
with Hume,
Kant, and
Maimon.

comes forward very early in Kant. It is to be found
in his *Einzig möglicher Beweisgrund, &c.* 1763. " The
conception of Positing or *Setzung* is perfectly simple,
and the same thing [*einerlei*] as that of Being in general." * * * " If we contemplate not the relation
(of positing) merely, but the thing posited in and
for itself, then this Being is equivalent to Existence
[*Daseyn*]."[1] The formal machinery of the *Critic of
Pure Reason* is a theory carrying out this notion.
Fichte adopted the term, *Setzung*, thought and word,
and his *Wissenschaftslehre* is nothing but the theory
of the positings, *Setzungen*, of the Absolute Ego, guided
by the purely logical principles of Contradiction and
Identity. Hegel generalised Fichte's Absolute Ego,
made it one and universal instead of many and individual, preserving the same method of purely logical
movement, the movement by Contradiction, as in
Fichte. This seems to me more nearly the " Secret
of Hegel" than anything which Mr. Stirling has told
us of it. It goes, for instance, more closely to the
root of the matter, than merely tracing the Kantian
Categories in Hegel's Logic, though that, too, is true
and valuable. It is through Fichte that Hegel's filiation to Kant must be traced.

The precise source of the peculiar post-Kantian
absolutism is to be found in Kant's identifying reflection with the Subject of reflection, instead of with
its object, the *distinction* which it *perceives* between
Subject and Object. Substance, agent, force, some
conditio existendi in short, was thus identified with
the Subject. This was the *Ding-an-sich*, the Absolute, and it mattered comparatively little whether it
was called Will, as in Schopenhauer, or Mind (*Geist*)

[1] Werke, Vol. I. p. 173. Ros. u. Sch. ed.

Book II.
Ch. IV.
——
§ 2.
Alignment
with Hume,
Kant, and
Maimon.

as in Hegel. The Subject was taken as an object of direct, instead of reflective, consciousness. It was taken as " the mind" in psychology may fairly be; a way not competent to philosophy. Philosophy and Psychology cannot on Kant's basis be distinguished; the latter is swallowed up by the former. What my theory does is to make possible a differentiation of the two, to mark off questions of nature and analysis for the one, questions of history and genesis for the other. The old strife " Psychologist or Ontologist" is ended, for in place of Ontology we have Metaphysic.

There were three fontal things in Kant, which it will be well to bring into a clear light. There was first his fontal error of thought,—besides his several errors of fact,—which consisted in confusing logical conditions, conditions *essendi*, with conditions of existence, *existendi*. This we have already seen in the case of apperception, the *Ich denke*. We know the pure or transcendental apperception only as a strain or thread in empirical apperception, namely, as the abstract fact of consciousness, in the whole or concrete moment of self-consciousness, the *dass* ich bin, or *dass* ich denke. This abstract thread or strain Kant raises to the rank of a prior condition of self-consciousness. The same mistake occurs again in the case of the Categories, and again in the case of Time and Space.

Secondly, this logical error is involved in the fontal thought which led Kant to his theory, and of which that theory is the expression, namely, that for all *necessity* there is required a transcendental condition in the conscious Subject.[1]

---

[1] See Kritik d. R.V. First edition. *Einleitung*, 2nd paragraph. Also *Einleitung* to the *Prolegomena &c.*

Book II.
Ch. IV.

§ 2.
Alignment
with Hume,
Kant, and
Maimon.

The third fontal thing in Kant is the Apperception itself, both pure and empirical. It is fontal as being the central point of the theory which he devised to express his fontal thought, involving his fontal error.

For Kant's hypothesis to account for the (supposed) necessity in certain cognitions, the hypothesis of a transcendental condition, I substitute an examination of the phenomena of primary consciousness. I go to the facts instead of an hypothesis. And with this result, as will I hope appear, that I find inseparable elements where Kant found *a priori* forms. This justifies a conclusion in many ways similar to Kant's, for inseparable elements, when inseparable from consciousness at large, as well as from each other in consciousness, are as necessary, being as universally present, as if they sprang from an original constitution of the Ego, or from an original constitution in the world, or in the Thing-in-itself. I am led by my analysis to justify the conceptions of necessity and its objective aspect, universality, though not by the same means as Kant did, nor yet in respect of quite the same phenomena.

The question is, whether any *a priori* furniture in the mind is requisite to account either for cognition at large, or for a class of cognitions which are supposed, or may turn out to be, necessary cognitions. And here again is seen the significance of my distinguishing and fixing upon the phenomena of primary consciousness, as the proper field of observation. Everything depends on what we take *as the data*, in proceeding to ask how cognition arises in, or is superadded to, the data. What states or processes of consciousness are to be held anterior to or below

Book II.
Ch. IV.
— —
§ 2.
Alignment
with Hume,
Kant, and
Maimon.

cognition? When you have fixed your data anterior to cognition, you can more readily see what is to be added to them, in order to bring out the result cognition. The distinguishing the data is itself a part, and possibly the most important part, of metaphysical analysis. For it may, and I think it will, be found, that these data comprise phenomena which are a sufficient account not only of the origin or formation of cognition, but also of the necessity in certain cognitions, so far as that necessity is an established fact.

It is in this question of the data that psychology, or rather the method proper to psychology but foreign to metaphysic, has done so much harm. I mean especially the rough classification of supposed faculties of the mind. The injury is not done so much by a theory of the mind as an entity, and of its faculties as separate powers or functions of that entity, as by the mere fact of the enumeration of separate functions, made mostly for some practical purpose, and then adopted as if the distinction were founded in the mental constitution itself. A mere enumeration of faculties or functions (it matters not which), such as Sense, Imagination, Judgment, Memory, Reason, Will, &c. &c., does violence to the facts of consciousness considered as mere phenomena, by forcing them into *a priori* classes, apart from their natural and actual relationships. We must learn to use all such terms as terms of denotation only, without assuming, tacitly and without proof, that they correspond to permanent divisions in the nature of the phenomena; we must use them as preliminary and provisional terms, as counters not as coin.

Kant is an arch-offender. Where is the justifica-

PRESENTATION AND REPRESENTATION. 235

Book II.
Ch. IV.

§ 2.
Alignment
with Hume,
Kant, and
Maimon.

tion for his saying dogmatically, at the outset of the Transcendental Logic: "Our cognition springs from two fundamental sources in the mind, the first of which is to receive *Vorstellungen* (Receptivity to impressions), the second the faculty of cognising an object through these *Vorstellungen* (Spontaneity of concepts); through the first an object is given to us" [observe the assumption that the " object" *exists*, and is given to us through the *Vorstellungen*] "through the second this object is *thought* in relation to this *Vorstellung* (as a mere *Bestimmung* of the mind)." * * * "Neither can these two faculties or capabilities exchange functions. The Understanding is incapable of intuiting anything, the Senses of thinking anything. Only from their combination can cognition arise."[1] The analysis and classification of faculties is thus made to give an independent support to the analysis and classification of mental states and processes which are their products; a support in this instance not warranted by the facts. A separate faculty of receptivity, and a separate faculty of spontaneity, are fictions.

The influence of this separation of faculties, one of which is spontaneity, is felt in the classification of what I have called their products, the various mental states and processes. In the list of these, the most general of all, says Kant, is *Vorstellung*, which he translates *repræsentatio*.[2] Everything comes under *Vorstellung* ;—perception, sensation, cognition, intuition, concept, and idea. Even sensation is a special mode, a differentiation, of it. But it is clear that

---

[1] Kritik d. R.V. 2nd edit. Einleitung to Transc. Logik, pp. 86. 87. Hartenstein, 1853.

[2] Id. id. p. 280.

BOOK II.
CH. IV.
———
§ 2.
Alignment
with Hume,
Kant, and
Maimon.

there must be a sense in which sensation, at least, must precede and be more general than *Vorstellung*, if *repræsentatio* gives Kant's meaning correctly. There must be a first instance before there can be a second, a *præsentatio* before a *repræsentatio*. There must be sensation as a *datum*, before there is sensation as a species of *Vorstellung*, that species of it which Kant defines in the above passage as "a perception" [i.e. a *Vorstellung mit Bewusstsein*] "which relates solely to the Subject as a modification of its state."

But from sensation as a *datum*, notwithstanding that it too is subject to the forms of time and space, as the ways in which alone we are sensibly affected, Kant turns systematically away. It does not lie in his road to examine it, convinced as he is from the first, that it will not conduce to prove his point of *a priori* forms of spontaneity. "The combination (*conjunctio*) of a manifold generally can never come to us through the senses, and therefore also cannot be contained along with [their content] in the pure form of sensible intuition; for" [mark the reason] "it is an act of spontaneity in the *Vorstellungskraft*."[1] Here is Kant's chief, and totally unwarranted, assumption. True, that combination which is an act of spontaneity (if there be such a thing) cannot come to us through the senses, or through their pure form; but there may be a "combination of a manifold" originally given in the sensation, and it is difficult to see how "a manifold," which is in the "form" of time, or in the "forms" of time and space, can be otherwise than a combined manifold. It would not be a manifold if it were not combined, seeing that it is in time,

---

[1] Kritik d. R.V. 2nd edit. § 15. p. 122. Hartenstein, 1853.

BOOK II.
CH. IV.

§ 2.
Alignment
with Hume,
Kant, and
Maimon.

or in time and space together. The form is the combination, the *together*, of the manifold.

Kant does not ignore this, but he systematically disregards it, and for the reason I have alleged. Thus we find him putting aside as empirical and contingent the phenomena of coexistence and sequence in sensations. "The transcendental unity of apperception is that unity through which all the manifold given in an intuition is united into a concept of the object. It is therefore called *objective*, and must be distinguished from the *subjective* unity of consciousness, which is a *determination* of the *inner sense*, through which that manifold of the intuition is empirically given to such a combination. Whether I can be conscious *empirically* of the manifold as together or in sequence, depends on circumstances or empirical conditions. Hence the empirical unity of consciousness, through association of *Vorstellungen*, itself concerns a phenomenon and is entirely contingent."[1]

And these are the terms in which he speaks of the "reproductive imagination" as he calls it, a few pages farther on: "So far, then, as the imagination is spontaneity, I call it also sometimes the *productive* imagination, distinguishing it thereby from the *reproductive*, the synthesis in which is subject solely to empirical laws, namely, those of association, and which therefore contributes nothing to the explanation of the possibility of *a priori* cognition, and on that account belongs not to transcendental philosophy, but to psychology."[2]

—— But to psychology! The first data of cognition, the simplest and most universal of all conscious

---

[1] Kritik d. R.V. 2nd edit. § 18. p. 127-8. Hartenstein, 1853.
[2] Id. id. § 24. p. 135.

Book II.
Ch. IV.
——
§ 2.
Alignment
with Hume,
Kant, and
Maimon.

states and conscious processes, belong not to transcendental philosophy, but to psychology!

" Hoc Ithacus relit, et magno mercentur Atridæ."

For note that by psychology Kant cannot mean his "rational" but his "empirical" psychology, not his "metaphysic of *thinking* Nature," but a part of "applied philosophy," a branch of knowledge which "must be wholly banished from metaphysic, and is wholly excluded therefrom by its very idea."[1] Could anything more plainly show the incompetence of the transcendental philosophy to lay down the bases of philosophy at large? Could anything more forcibly bring home the necessity of instituting an enquiry on broader principles, an enquiry which shall embrace all the phenomena of consciousness, all its states and processes, from the least to the greatest; above all an enquiry that shall not be directed to establish a foregone conclusion, or support a favourite hypothesis?

Neither my position nor Kant's can be fully understood without understanding that of an earlier philosopher, who was the determining condition of Kant, I mean Hume. Hume was no purposed sceptic, but his reasoning led him into contradictions which made all philosophy impossible. No one can insist on this more forcibly than Hume himself, in the conclusion of the First Book of the Treatise of Human Nature. He sees the contradictions, he sees also a necessity of falling into them. There must be something radically unsound in self-contradictory positions of this kind. Whether the error can be discovered, and when discovered removed by logic, and a true theory established instead, is another matter.

[1] Kritik d. R.V. 2nd edit. Methodenlehre, p. 600. Hartenstein, 1853

Book II.
Ch. IV.
§ 2.
Alignment
with Hume,
Kant, and
Maimon.

The character of Hume's writings is very peculiar. In some respects he is a true metaphysician, and his method is the method of reflection. It is so as far as the analysis of aspects goes. Phenomena are with him everything. There is no Thing-in-itself in Hume. Impressions and Ideas are existence; existence, objectively, means impressions and ideas. He belongs to the English school of philosophy, and is the immediate successor of Locke and Berkeley. He treats all existence on the very same principles on which Berkeley treated "matter." He asks on every occasion, What do we *mean* by such and such words and phrases? He goes to reflective consciousness for the answer.

But there his reflective method ends. He does not perceive that he is adopting the reflective method, nor what its nature is, and consequently he does not push it far enough, does not make all the use of it possible. It dominates him, not he it; and there soon comes a point where he lets it escape him, and falls back into the direct method. That point is where he comes to what I call the analysis of elements, the analysis which we are entering on in the present Chapter. Here it is that Hume ceases to be a metaphysician and becomes a separatist; here it is that he mixes up his scientist with his metaphysical theories; and this mixture is the substance of his scepticism. Let us see how this is brought about.

There are three fundamental conceptions which lie at the basis of Hume's philosophy of the Understanding:

1st. The dependence of Ideas on Impressions; or, as I should say, of representations on

Book II.
Ch. IV.

§ 2.
Alignment
with Hume,
Kant, and
Maimon.

presentations. This is the basis of his phenomenalism; this is what is truly reflective and metaphysical.[1]

2nd. The indistinguishable or atomic character of simple impressions and ideas. This is an assumption; and this it is which, in combination with the third principle, leads to his sceptical results, and constitutes what I may call his subjective atomism.[2]

3rd. The doctrine that whatever is distinguishable is also separable, and whatever is separable is also distinguishable.[3]

Hume's assumption is, that impressions come to us originally in separate indivisible drops, so to speak; in other words, he assumes them to be what I call objects of *direct* perception. " Simple perceptions or impressions and ideas are such as admit of no distinction nor separation. The complex are the contrary to these, and may be distinguished into parts."[4] And again, " Every simple idea has a simple impression, which resembles it, and every simple impression a correspondent idea."[4] And what he thus finds separate he keeps so, by means of his third principle, the separability of whatever can, and consequently the inseparability of whatever can not, be distinguished. For the third principle is nothing

---

[1] Treatise of Human Nature. Book I. Of the Understanding. 1739. In Messrs. Green and Grose's edition of 1874, Vol. I. pp. 312. 370, 375 385.

[2] Id. id. Vol. I. pp. 337 8. 344. 345. 349. 375. 403-4. 436. 450. 463. 474-5. 517-8. 528.

[3] Id. id. Vol. I. pp. 326. 332. 335. 343. 344 359. 518.

[4] Id. id. Vol. I. pp. 312. 313.

Book II.
Ch. IV.

§ 2.
Alignment
with Hume,
Kant, and
Maimon.

more than insisting on the significance of the supposed fact stated by the second. If it should turn
out on further analysis, that there are no such things
as simple perceptions in which no distinctions are
found, but that every percept, however near it may
approach simplicity, is never perfectly simple, then
the doctrine founded on this supposed fact will fall
to the ground; for there will be discovered cases of
distinguishable things which are nevertheless inseparable. It may turn out that Hume is right after
all; but it requires a further analysis than he gives
us to show it.

In conjunction with these two principles, his first
principle, however sound in itself, goes but little way
towards a system of philosophy. It does but show
the need for one, and bring the problem out in its
true difficulty. We shall see this plainly from the
following passage: "When any object is presented
to us, it immediately conveys to the mind a lively
idea of that object, which is usually found to attend
it; and this determination of the mind forms the
necessary connexion of these objects. But when we
change the point of view, from the objects to the perceptions; in that case the impression is to be considered as the cause, and the lively idea as the effect;
and their necessary connexion is that new determination, which we feel to pass from the idea of the
one to that of the other. The uniting principle among
our internal perceptions is as unintelligible as that
among external objects, and is not known to us any
other way than by experience. Now the nature and
effects of experience have been already sufficiently
examin'd and explain'd. It never gives us any insight into the internal structure or operating prin-

Book II.
Ch. IV.

§ 2.
Alignment
with Hume,
Kant, and
Maimon.

ciple of objects, but only accustoms the mind to pass from one to another."[1]

There is the point upon which Kant joined issue. Both Kant and Hume adopt the subjective and reflective method; but Hume appeals to experience, and discovers no causal or constitutive nexus between phenomena ; Kant discovers, in the transcendental apperception as we have seen, a synthetic spontaneous power which constitutes cognition and is the condition of experience itself. Such is the issue; and to me it seems that, notwithstanding Kant's analysis goes far deeper than Hume's, the issue itself is not thereby decided. In the end Kant makes the assumption of an entity, or a force, which Hume has already, by anticipation, shown to be a derivative of experience, and therefore not a legitimate explanation of it. "All ideas are derived from, and represent impressions. We never have any impression, that contains any power or efficacy. We never therefore have any idea of power."[2]

Kant in fact overshot the mark. Yet at the first outset of the Critic, in its later form, in the Introduction to the *Second Edition*, there is a passage in which he is within an ace of discovering where the true knot of the question lay. All our cognition springs from, and must be based upon, experience, said Hume. No, replies Kant; all our cognition begins with experience, but it does not all spring out of it, and therefore is not entirely based upon it. In the corresponding passage of the *First Edition* he

---

[1] Treatise of Human Nature. Book I. Of the Understanding. 1739. In Messrs. Green and Grose's edition of 1874, Vol. I. p. 463. § On the idea of necessary connexion.

[2] Id. id. Vol. I. p. 455. § On the idea of necessary connexion.

Book II.
Ch. IV.

§ 2.
Alignment
with Hume,
Kant, and
Maimon.

had said, "Experience is undoubtedly the first product brought forth by our Understanding, in working up the raw material of sensible impressions." One
only wonders how Kant either missed the discrepancy
between his own meaning of the term experience and
Hume's, or, seeing it, how he allowed its significance
to escape him. The discrepancy is this. Hume
meant by experience all our impressions and ideas
without distinction, all the content of consciousness.
Kant meant by it our *ordered* experience as we find
it when we begin to philosophise, the ordered experience of the ordinary understanding. Hume's use
of the term was much the larger. But both uses
tacitly involved the assumption that there was, as a
fact, a portion of experience in the larger sense, which
was chaotic and not orderly, instead of simply raising
the question whether this was the fact or not.

True, Hume's larger use of the term gave him
this advantage, that it justified the answer which I
have said above he might have made to Kant. For
nothing could possibly lie beyond experience, *jenseit
der Erfahrung,* in Hume's large sense of the term.
At the same time Hume's further assumption, of the
separateness of impressions, was itself a narrowing of
his large conception of experience, and tantamount
to an assumption concerning its orderliness.

The true objection to Hume should have run
somewhat as follows:—You are really including an
orderliness in your term experience, and you are
assuming that it is a contingent orderliness *ab extra,*
though you think you are making no assumption at
all. You are only entitled to assume experience,
undetermined as to whether it is chaotic or orderly.
Orderliness is a feature which requires accounting

Book II.
Ch. IV.

§ 2.
Alignment
with Hume,
Kant, and
Maimon.

for. You must analyse experience without assuming
either an original orderliness or an original chaos,
and then you will possibly see in what its orderliness
consists, as we find it when we begin to philosophise.
That orderly experience may have a source beyond
itself, and yet not beyond experience at large. You
must see whether experience is ever chaotic, ever
not-orderly.—The question between a transcendental
and a phenomenal source of real cognitions, as well
as between their necessary and their contingent character, would thus have been fairly raised. For, if
it turned out that experience in the large sense was
always and in all its parts orderly, and never chaotic,
then the source of necessary cognitions would have
been found in that orderliness, within and not beyond
experience. But Kant, instead of making this kind
of objection to Hume, made an opposite kind of
assumption. Whereas Hume had assumed that experience, in the large sense, had a contingent orderliness, Kant assumed that a portion of it, which he
excluded from experience in his sense, was originally
chaotic, and then he derived its orderliness (where it
existed) from a transcendental source.

Hume's assumption of the separateness of impressions prevented him from finding any real nexus
in the supposed nexus of experience. He began with
ordered experience as a given fact, and this dissolved
its apparent necessity. Kant on the other hand
assumed a transcendental principle expressly to account for ordered experience and its necessity; but
this principle was really a product of that ordered
experience which it was called in to account for.
The power, which it assumed as the condition, was
really the product of ordered experience. Nothing

Book II.
Ch. IV.

§ 2.
Alignment
with Hume,
Kant, and
Maimon.

guaranteed the existence of the power, unless it could be found in ordered experience itself. And this it could not be so long as Hume's reply could be justified by his large use of the term experience, and was not invalidated by discovering the fallacy of his analysis.

Salomon Maimon is the only philosopher, so far as I know, who has seen the problems of philosophy from the same point of view as Hume and Kant. He aligns himself with them, so to speak. The simple straight-forwardness which enabled him to take this view, when all the world around him was carried away with the Kantian torrent, marks him in my estimation as the greatest philosopher since Kant. His mind was steadily fixed on the single purpose of philosophical truth. He thus saw not only what Kant had said and done, but what had conditioned that saying and doing. Germany, as if actuated by an implicit belief that Philosophy is of the Germans, soon began to regard its Kant like a revelation from heaven. And this they are still doing. "Go back to Kant," is their cry; "*Kind, bet' ein Vater-unser.*" Good; but not enough. They must go back to Hume as well. As it is, they align themselves with Kant and Leibniz, not with Kant and Hume. This is why I said in the first Chapter, that a greater philosophy could arise in the line of Locke, than can ever arise in the line of Leibniz. The post-Kantian philosophy of the absolute, in all its branches, is a mere episode in the general history, a side current from the main stream, of philosophical thought.

Not the great and famous of the Universities, not the theologians, not the state-philosophers, not the acute psychologists, not the imaginative ontologists, but Maimon the Polish Jew, with his unobtrusive

Book II.
Ch. IV.
§ 2.
Alignment
with Hume,
Kant, and
Maimon.

analytic genius, has been the real hander on of the torch of truth.

I say this without meaning to assert that Maimon reached a position which was theoretically satisfactory. He characterises his own philosophy and method as "sceptic-critic." In fact, he carries on the critical strain in Kant, just as Hume carried on the critical strain in Berkeley. Maimon's tendency therefore is to redress the balance, to turn it in Hume's direction. He retains the conception of the Transcendental, and he retains the Kantian Categories, but excludes causality from the list. Real cognition, according to Maimon, is confined to mathematical objects, not extended to sequences of empirical objects in time. Everything is excluded from real and necessary cognition, which does not fall under the *Satz der Bestimmbarkeit*, (Determinability), which is the central principle of Maimon's system. He was here no doubt on a true track. My only wonder is that he did not see how much he had included in real cognition; that, in including mathematical objects, he had virtually left nothing excluded; that he still continued to call himself a sceptic at all, though not a Humian sceptic.[1]

I follow, then, in the train of Maimon, and like him I ask *What are the facts?* We have seen that Hume, starting from the assumption that percepts are from the first separate from one another, arrives at the conclusion that there can be no necessary connection between them in thought. Thought is, with

---

[1] See Maimon's *Versuch einer neuen Logik*, Berlin, 1798, 2nd edit. p. 190-3. for his "sceptic-critic" position. Also a general summing up of his views in the *Briefe an Œnesidemus*, in the same vol. p. 322 et seqq. and again at p. 425-438.

him, a particular *de facto* process, like eating and sleeping. It will perhaps be said—'What would you have more? Is not perception itself a *de facto* process? Is necessity higher than fact? The conception is impossible; necessity itself has no other than a *de facto* existence.' Most true. But what we want *more* in thought, than what Hume allows to it, is, that it shall not *lose* any of the necessity which belongs to perception as a *de facto* existence. We want to have thought shown to be not a particular but an universal process; not a contingent process, but one equally universal and equally necessary with perception. We want to have the nexus of thought between percepts made, in its nature, as indissoluble as that which holds a single percept together. Certainty is a thing which has two terms. The nexus between them must be as real as the terms themselves. That is the condition of escaping from scepticism.

BOOK II.
CH. IV.
§ 2.
Alignment
with Hume,
Kant, and
Maimon.

We have seen Kant, in order to escape from the scepticism into which Hume was led, endeavouring to secure this nexus by the fiction of a transcendental constitution of the thought-faculty, a power of synthetic combination which was the prior condition, not only of framing a nexus between percepts, but of framing percepts themselves. Hume's "impressions," as well as his sequences of impressions and ideas, were to be made, made originally, by this power. The *de facto* trains of association which pass through consciousness, which according to Hume were the basis and condition of reasoning, were now to be themselves referred to an *a priori* source. The particular *de facto* process was again to become a necessary one, in virtue of its origin in the transcendental apperception.

Book II.
Ch. IV.

§ 2.
Alignment
with Hum ,
Kant, and
Maimon.

Now this was taking the seat of the symptom for the seat of the disease. The symptom was weakness in the combining power of *thought*. Then, argued Kant, let us strengthen the thought-faculty; Hume has overlooked the circumstance that a faculty of synthesis has presided at the construction of his "impressions," which he takes to represent objects. The thought-faculty is the object-making faculty.

But let us go to the facts. Perhaps it will appear from the analysis of Hume's "impressions," from the analysis of percepts, that the nexus which binds percepts together in thought and the nexus which holds a single percept together are one and the same; that the former is only the latter modified; that no thought-faculty need be assumed, in order to give to the process of thought the same necessity and universality as perception itself possesses. The analysis of perception must decide.

§ 3. I enter now, without further prologue, upon the promised analysis of the phenomena of primary consciousness. I go straight to the facts, without saying I go to perception, or sensation, or thought, or any special mode at all. What I find, when I look at consciousness at all, is, that what I cannot divest myself of, or not have in consciousness, if I have consciousness at all, is a sequence of different feelings. I may shut my eyes and keep perfectly still, and try not to contribute anything of my own will; but whether I think, or do not think, whether I perceive external things or not, I always have a succession of different feelings. Anything else that I may

BOOK II.
CH. IV.
——
§ 3.
Analysis of
minima of con-
sciousness.

have also, of a more special character, comes in as part of this succession. Not to have the succession of different feelings is not to be conscious at all.

This fact, the succession of different feelings just described, is what I am going to examine. I use the names describing it purely *denotatively*, to designate the phenomenon which I mean to have before me. And now I add, that this phenomenon seems big enough for the first section of the enquiry, and for the present Chapter. It contains the *data* simply. What I am *not* going to include in it is the modification of this succession by a conscious act, what Kant called spontaneity. That will come in the next Chapter. There is a great deal included in the large term, succession of different feelings, which we shall have to bring out into distinctness. This will of course require acts of conscious thought on our part; it has already required one to mark out the object itself. But acts of this kind are not to be now examined, are not to make part of our present object-matter. I am going to state one by one what features I find in those phenomena which are described by the words *a succession of different feelings*, without taking account of other features in them which are not described by those words. And I shall first attempt to do this by taking a supposed single case, artificially isolated, and reduced to its lowest terms, a *minimum* of consciousness. Then will follow a second part of the analysis, in which the succession will be treated in portions of considerable length. First, then, for the analysis of *minima*.

1. The first point to be noticed in this phenomenon is the difference of the feelings. The minimum of consciousness contains two different feelings. One

Book II.
Ch. IV.
———
§ 3.
Analysis of
minima of con-
sciousness.

alone would not be felt. The second brings the first into consciousness together with itself. Still it is not a case of what is sometimes called " latent consciousness," not a case where the first of the two feelings is below the threshold of consciousness until followed by the second. Neither of them is distinctly in consciousness before the other. But of this apparent simultaneity there are two cases: the first is that of a real simultaneity, the two sub-feelings are really parts in coexistence, not in succession; the second is that in which one of them is felt as growing fainter (called *going* when referred to its place in succession), the other as growing stronger (called *coming* when referred to the succession). The simultaneous perception of both sub-feelings, whether as parts of a coexistence or of a sequence, is the total feeling, the minimum of consciousness, and this minimum has duration. There is duration not only in the case where the terminus a quo and the terminus ad quem together, or in other words the change from this to that including both, is the minimum of consciousness; but also in the case of coexistence of the sub-feelings as simultaneous portions of the whole minimum.

Looking at the minimum of sequence alone and isolated, we should not know that what we call the first of the two sub-feelings came first in order of time at the moment of consciousness; we know this afterwards, when we have had many such changes, forming parts of one continued train. Time-duration, however, is inseparable from the minimum, notwithstanding that, in an isolated moment, we could not tell which part of it came first, which last. In both cases of minima alike, those of coexistence and those of sequence, one sub-feeling is as it were relieved against the other, and this relief it is which brings both into consciousness.

Whatever is paradoxical in this description arises from the necessary use of language, which is all formed of words of second intention, to describe phenomena in their first intention. We have to correct one statement by its apparent opposite, in order to convey a picture of states of consciousness existing before those distinctions, which our statements involve, were drawn or perceived. We have no other means of depicting the earliest states of primary consciousness, and the same holds good for depicting states of consciousness artificially isolated for the purpose of analysis, the minima of consciousness.

We do not require to know what difference is, or what similarity is, in order to be sensible of difference or change of feeling. We feel the *thing*, without knowing the *thought*, or *name*, of the thing. Or to put the same thing in technical language,—we know difference or change in its *first intention*, but not in its *second*. And just the same may be said of the time sequence involved in the minimum of consciousness. We do not require to know that the sub-feelings come in sequence, first one, then the other; nor to know what coming in sequence means. But we have, in any artificially isolated minimum of consciousness, the *rudiments* of the perception of former and latter in time, in the sub-feeling that grows fainter, and the sub-feeling that grows stronger, and the change between them.

And therefore it cannot be argued that, because in a minimum of consciousness, taken alone, there is no distinct perception of former and latter, (if there

Book II.
Ch. IV.

§ 3.
Analysis of
minima of con-
sciousness.

were, it would be no minimum), therefore the perception of sequence is not a *datum*, but an addition made by imagination or inference. For the minima of consciousness are not *data* as minima, nor does consciousness come to us in distinct minima originally; but we it is who reduce it to minima artificially. And the very distinction of a minimum, so made, cuts it off from the continuous stream which it is part of, treats it *statically*, and thereby involves the assumption that its sub-parts are simultaneous. But this simultaneity of sub-parts, being introduced by our method, must not be reckoned to the thing analysed. We must keep that clear of the assumption due to this source. And then we find, what alone we can expect to find in a minimum, that there are in it the *rudiments* of former and latter states of consciousness, in its sub-feelings.

2. In the next place I remark that the rudiments of memory are involved in the minimum of consciousness. The first beginnings of it appear in that minimum, just as the first beginnings of perception do. As each member of the change or difference, which goes to compose that minimum, is the rudiment of a single perception, so the priority of one member to the other, although both are given to consciousness in one empirical present moment, is the rudiment of memory. The fact, that the minimum of consciousness is difference or change in feelings, is the ultimate explanation of memory as well as of single perceptions. A former and a latter are included in the minimum of consciousness; and this is what is meant by saying that all consciousness is in the form of *time*, or that time is the form of feeling, the form of sensibility.

PRESENTATION AND REPRESENTATION. 253

Book II.
Ch. IV.
___
§ 3.
Analysis of
minima of con-
sciousness.

Crudely and popularly we divide the course of time into Past, Present, and Future; but, strictly speaking, there is no Present; it is composed of Past and Future divided by an indivisible point or instant. That instant, or time-point, is the strict *present*. What we call loosely the Present is an empirical portion of the course of time, containing at least the minimum of consciousness, in which the instant of change is the present time-point. The stimulus or shock given to the nerve, or the change between two intensities of stimulus, which is the psychological condition of the feeling, corresponds to the instant of change between the two sub-feelings constituting the minimum; and if we take this as the present time-point, it is clear that the minimum of feeling contains two portions, a sub-feeling that goes and a sub-feeling that comes. One is remembered, the other imagined. The limits of both are indefinite, at beginning and end of the minimum, and ready to melt into other minima, proceeding from other stimuli.

Time and consciousness do not come to us ready marked out into minima; we have to do that by reflection, asking ourselves, What is the least empirical moment of consciousness? That least empirical moment is what we usually call the present moment; and even this is too minute for ordinary use; the present moment is often extended practically to a few seconds, or even minutes; beyond which, we specify what length of time we mean, as the present hour, or day, or year, or century.

But this popular way of thinking imposes itself on great numbers even of philosophically minded people, and they talk about the *present* as if it was a *datum*, as if time came to us marked into present

Book II.
Ch. IV.

§ 3.
nalysis of
ima of con-
:iousness.

periods like a measuring tape. It is an instance of the separatist fallacy of using direct instead of reflective methods in subjective analysis; present periods are taken as so many separate objects, simply because we are accustomed to mark time off into past, present, and future, for practical purposes. And when this is done, then some hypothesis is inevitably required, to explain how what is so sundered is to be re-united, in order to conform to facts.

It is thus for instance that Mr. Ward finds it necessary to assume a faculty of Intuition, in order to explain how memory can be trustworthy; in other words, he endows us with a faculty of memory, a faculty capable of passing intuitive judgments which carry with them their own evidence of truth. I refer to his argument against the "experience" school of philosophers, selecting Mr. J. S. Mill as its representative, in the Philosophical Introduction to his work on Nature and Grace.[1] In that argument Mr. Ward says: "You make use of your own past experience,—you make use of other men's experience,—as part of the foundation on which you build. How can you even guess what your past experience has been? By trusting memory. But how do you prove that those various intuitive judgments, which we call acts of memory, *can* rightly be trusted? So far from this being provable by past experience, it must be in each case *assumed* and *taken for granted*, before you can have any cognizance whatever of your past experience." And in farther testing a case of memory we find: "The question which I would earnestly beg Mr. Mill to ask himself is this;—what is my ground for believing that I *was* cold a short time ago? 'I

---

[1] On Nature and Grace, London, 1860, p. 26.

PRESENTATION AND REPRESENTATION. 255

Book II.
Ch. IV.
§ 3.
Analysis of
minima of con-
sciousness.

have the present *impression* of having been cold a short time ago;'—this is one judgment. 'I *was* cold a short time ago;'—this is a totally distinct and separate judgment. There is no necessary, nor even probable, connexion between these two judgments,—no ground whatever for thinking that the truth of one follows from the truth of the other,—except upon the hypothesis, that my mind is so constituted as accurately to represent past facts. But how will either 'sensation' or 'consciousness,' or the two combined, in any way suffice for the establishment of any such proposition?"[1]

The answer is easy. Sensation and consciousness will not suffice, when an arbitrary and unphilosophical separation is made between past and present; they will suffice, as future analysis in the proper place[2] will show, when they are examined by the method of reflection.

3. It will be noticed that I have employed the word Perception. I mean to use that term to express all that we have now found to be involved in the succession of different feelings, as it has now been described; I use it to signify any empirical portion of this train of conscious states, whether that portion is a minimum of consciousness, or larger than a minimum, or the whole train. But I do not include in it any means by which one portion of the train is marked off from another. I assume that this can be done, that we are familiar with its being done, and that we can contrast the train in which it is not done with the train in which it is done. If we suppose, for argument's sake, that the train is marked off into

[1] On Nature and Grace, London, 1860, p. 28.
[2] In the present Chapter, p. 274.

Book II.
Ch. IV.
———
§ 3.
Analysis of
minima of con-
sciousness.

distinct portions, it is *we* who so mark it off; and I abstract from this circumstance of its being so distinguished into determinate portions in calling the whole —perception, or a train of percepts.

Perception means one thing in metaphysic, at least as I now define it, and another thing in psychology. In psychology it means sometimes a definite, marked off, portion of the train of sensations together with the reason for marking them off as separate, i.e. together with their *nexus inter se;* and sometimes it means a portion of the train of sensations so marked off *and referred to an object* external to the mind or train of sensations. In the first of these two senses, perceptions pre-suppose a process of thought, by which the nexus is discovered or supplied; in the second, they pre-suppose also that reflection has taken place, so as to distinguish *objects* from the subjective train of sensation.

It is in the second sense that the term perception is employed by Professor Delbœuf, in a recent work.[1] "Si de plus l'être est *connaissant,* s'il est doué d'*intelligence,* ce mot étant pris dans le sens le plus étendu qu'il puisse avoir de manière à s'appliquer aux animaux même inférieurs, il aura des *perceptions,* c'est-à-dire qu'il rapportera sa sensation à une cause en général autre que lui, et qu'il attribuera à cette cause une qualité, qui sera celle de lui procurer une sensation déterminée."

This use of terms is perfectly legitimate in psychology, and the best adapted for its purposes. For psychology begins with its whole object-matter already divided into conscious beings and their environment, and enquires into the relations which obtain between

[1] Théorie générale de la Sensibilité, 1876, p. 5.

BOOK II.
CH. IV.

§ 3.
Analysis of
minima of con-
sciousness.

the two divisions. It aims too at quantification, or measurement, of the stimuli from without the organism, of the reaction of the organism from within, of the intensity and duration of the conscious states, and of the relations of these to each other. It is best for it, then, to take sensations as single states of consciousness, not distinguished into inseparable elements; and next to go to perceptions, as that kind or that case of sensations, where the notion of stimulus, or object which is the source of the stimulus, becomes part of a group of sensations. The train of sensations is a *datum* both in metaphysic and in psychology; but metaphysic treats it in one way, because its aim is analysis of conscious states; and psychology in another way, because its aim is the determination and measurement of the conditions which produce and modify it. When psychology analyses states of consciousness simply, it is in subordination to that method.

Thus, to take another instance, Herr Professor Wundt[1] devotes one whole section to the *Empfindungen;* the next to the *Vorstellungen;* the next to *Bewusstsein* and interaction of *Vorstellungen;* the next and last to the *Bewegungen.* And in justifying this procedure he says: "In beginning the consideration of the inner processes with their simple phenomena, we find ourselves constrained at the outset to confess that the Simple itself never offers itself to our observation, but must be always first separated from the comparatively intricate relations which it enters. Those psychological elements, which indubitably bear the character of being the most simple phenomena, are

---

[1] Grundzüge der Physiologischen Psychologie, 1874, p. 273. See also p. 365 and 522-3, as to sensations of sight.

the pure *Empfindungen*. We mean by them the most original states which the human being finds in himself, separated from all references and relations which the developed consciousness always brings about. Conceived in this abstract way, *Empfindung* possesses solely and alone *Intensity* and *Quality*, as closer determinations. On the other hand its time duration remains, at first, out of our consideration, because the time-intuition first develops itself in the interchange of *Empfindungen* and *Vorstellungen*. Similarly we abstract wholly and entirely from the space relations, in which certain *Empfindungen* are always given to our self-observation, because these relations, as will be seen later on, always arise out of the interaction of a plurality of *Empfindungen*. Pure *Empfindung* so defined is therefore nothing more than an inner existence, variable in strength and quality, which we may perhaps make clear to ourselves by the fiction of a Condillac statue, supposed to be just entering on activity."

On this I remark, that the method of treatment, the method of abstracting purposely at first from everything but quality and intensity in feelings, puts out of view the question whether as a matter of fact time duration is an original element in those feelings, and contained in their minima; for the " time-intuition" (*Zeitanschauung*) may very well not only develop itself but even originate in the interchange of *Empfindungen* and *Vorstellungen*, and yet duration be an element in the minimum of sensation. The former question depends on how you define "time-intuition." Wundt takes[1] *Wahrnehmungen* and *Anschauungen* as two names, (the first with an objective, the second

---

[1] *Grundzüge der Physiologischen Psychologie*, 1874, p. 461.

Book II. ·
Ch. IV.

§ 3.
Analysis of
minima of con-
sciousness.

with a subjective reference), of those *Vorstellungen* which have real objects. And *Vorstellung* he takes to mean a picture (*Bild*) of an object produced in consciousness. So that time-intuition must mean with him a time-picture produced in consciousness. And a time-picture no doubt must originate in consequence of a considerable number of single sensations.

But the question remains, whether or not these single sensations contain perceived duration in them. If they do not, whence comes time duration into consciousness? If they are not the source, what is? I can imagine only three answers;—either an *a priori* form in the mind; or an otherwise unknown constitution of things-in-themselves; or resting content with the fact as an ultimate fact of experience. Now Wundt puts together percepts out of sensations; a perception is a whole of which sensations are the parts. "Die Vorstellung ist im Vergleich mit der Empfindung ein Zusammengesetztes. Sie enthält Empfindungen als ihre Bestandtheile."[1] He adopts then the first of the two definitions of perception mentioned at p. 256, as Delbœuf adopts the second.

How then does Wundt explain to us the composition of perceptions out of sensations? " This composition," he tells us at the same place, " can proceed in a double manner; first in the form of an arrangement of sequence in time (*einer zeitlichen Aneinander-reihung*), and secondly as an ordering in space (*als eine räumliche Ordnung*). Both combinations rest upon special applications of the general law of relation." But can any one frame a notion of a *law of relation* which does not presuppose duration, se-

[1] Grundzüge der Physiologischen Psychologie, 1874, p. 465.

Book II.
Ch. IV.
———
§ 3.
Analysis of
minima of con-
sciousness.

quence, and coexistence? If it does so, it cannot be called in to account for them. And yet Wundt proceeds: "Where the sensations combine themselves in the time form, succession and simultaneity are the result (*ergeben sich*) as the essential distinctions of the perception." This "general law of relation," as an explanation of the existence of time and space relations, seems to me to have no more justification than Mr. Ward's "intuitive memory."

I conclude, then, that the psychological method has led Professor Wundt so to isolate the sensations as to render their combination into perceptions inexplicable, unless it were by recourse to a transcendental thought faculty, so long as we continue in that method. He has made a cleft between sensations themselves, and between sensations and perceptions, which he offers no means of bridging. There is nothing but the statement that it must be bridged, because it is not found in experience. For this is what the following passage seems to amount to: "*Pure* sensation (*Empfindung*) is an abstraction which never comes forward in our consciousness. Consciousness possesses *only* perceptions (*Vorstellungen*): the sensations are in it always arranged according to the general forms of intuition, time and space. Nevertheless we are compelled by an overwhelming number of psychological facts, which have been explained in the preceding section, to suppose the existence of pure sensation, and to assume that perceptions everywhere form themselves, by a psychological synthesis, out of sensations."[1] I am not convinced of the necessity, because I find sensations always combined with rudiments of time or of

[1] Grundzüge der Physiologischen Psychologie, 1871, p. 711-12.

BOOK II.
CH. IV.
——
§ 3.
Analysis of
minima of con-
sciousness.

time and space, which I call the formal element. These rudiments it is from which they are never found pure. A state previous to perception, in which sensations exist *pure*, is a fiction.

4. I next come to the feature in the succession of different feelings which furnishes the title to the present Chapter, presentation and representation. We have the train of conscious states passing before us, through a time-point which marks the centre of that portion which we call the empirical present moment. We may figure it to ourselves by walking along a pavement, and watching the slabs seeming to meet us and pass away under our feet behind us.

Every state called a presentation is an empirical present; but not every empirical present is a presentation. The content of an empirical present may be wholly representation, consisting wholly of old, recalled, or re-moulded, presentations. It may also consist of wholly new presentations. And it may also consist partly of one and partly of the other; and this is by far the commonest case. It is moreover very difficult, often impossible, to separate, in an empirical present moment of mixed presentation and representation, what is new and what is old. Each impression on the senses as it comes is new; but the framework into which it falls, the figures of things, are old; and how much precisely belongs to the one, how much to the other? The only purely subjective criterion between a new and an old content of perception, the only criterion, that is, which does not rest on prior extraneous knowledge of the source of the sensations, is, I believe, their *vividness*. There is a certain vividness or intensity about what we actually receive through the senses, which is

Book II.
Ch. IV.

§ 3.
Analysis of
minima of con-
sciousness.

wanting to what we re-receive through memory or imagination.

But, judged by this test, presentation is not confined to sensations. Emotions also are presentations sometimes. There is a peculiar vividness also in the imagery of some dreams, which seems to mark it as new and not a mere re-moulding. Besides this there is often in dreams an emotional vividness, one attaching to the emotionally pleasureable or painful character of their imagery, a freshness in its contained feeling, as if the cerebral machinery which reproduced it were acting as stimulus to a distinct organ of internal and emotional sensibility.

The distinction between presentations and representations, then, does not lie solely in the circumstance of presentations being present to consciousness for the first time, that is, of being new and not old. This alone gives no distinction of kind. Every state of consciousness is different in some respect from every other; and in that respect will be new and not old. A state of consciousness must have some other mark, besides that of being a present moment and in some respect new, in order to be classed as a presentation. It does not get this other mark simply from the element of feeling (sensation or emotion) which it contains; for representations contain this element of feeling, and cannot exist without it. But the criterion of presentations is drawn from a particular class or kind of feeling,—the feeling of vividness which accompanies some states of consciousness and not others, a peculiar feeling not confined to sensation, but attaching also to emotion. I do not say that this vividness renders the emotions and dream images to which it attaches indistinguishable from

PRESENTATION AND REPRESENTATION. 263

BOOK II.
CH. IV.

§ 3.
Analysis of
minima of con-
sciousness.

presented sensations; far from it. But I say that they are alike in possessing one peculiar feature, vividness, which is the sole purely subjective criterion of presentations.

That newness, freshness, and vividness are not confined to sensations alone, is shown by the case of ordinary rhetoric, when we vary our expressions in order to heighten their effect. We stimulate the imaginative emotion by exhibiting the mental picture in new lights, just as we examine an object of sight by changing the incidence of its rays on the retina, or as we play a passage of music over again in a different octave. In all these cases alike we obtain new and fresh feeling, by bringing the nervous organ of feeling into contact with the object in a new way, by setting up new changes in it, calling untired portions of it into action. But the kind of feeling obtained remains the same, is old and not new.

And now as to the bearing of this distinction. Its importance is of the highest order, for it is presentations alone that afford the ultimate verification for all opinions and hypotheses. The *newness* of presentations gives them rank as an independent source of truth. Newness not in kind, but in time. Oldness of kind and newness in time,—that is, in other words, a repetition, a fresh instance, a verification of a remembrance. A reversal of the order of first experience, from presentation to representation; a reversal which repeats the process backwards from representation to presentation.

True, it is a distinction between empirical moments of time, not between inseparable elements; for a presentation is an empirical portion of the train of consciousness. But this does not detract from, on

Book II.
Ch. IV.
———
§ 3.
Analysis of
minima of con-
sciousness.

the contrary it adds to, its practical importance as a test. It may be itself justified by analysis, but in order to be employed it must be empirical and immediately available.

To see farther the bearing of this distinction, we must have recourse to the psychological distinction between the organ and its environment. We then may divide presentations into those of external and those of internal sensibility, and to the latter will belong all those cases where remembered or imagined feelings are reproduced with the special vividness which we have seen to be the subjective criterion of presentations. Reflection itself is a case of this branch of presentation. It is a case of presentation arising in redintegration,—redintegration being the name for the train of consciousness as a train, and so far as it is not interrupted by feelings coming from without the nerve organism. To the fact of its being presentation reflection owes its verifying power; its character of a court of final appeal in all cases of interpreting facts. If the presentations of perception, including those of emotion as well as those of sense, are the first source and final verification of facts simply, those of reflection, which is thought as well as perception, are the final verification of what those facts are, what they consist of, what they *mean*. The one may be compared to acts of parliament, the other to judicial decisions interpreting them.

It must not be forgotten that presentation alone is not verification. More than that is required for verification. The fact again presented must be constant, or at least capable of being presented repeatedly. Oldness of kind, with newness of time; repetition of the instance, and as often as doubt may arise con-

Book II.
Ch. IV.

§ 3.
Analysis of
minima of con-
sciousness.

cerning it;—that is at least the ideal of verifiability. Emotions may be presentations; dreams may be presentations; but the question is—are they renewable, are they renewable at pleasure? No. Then they are not verifications.

5. It is perhaps hardly necessary to state formally, what has been assumed all along, that the succession of different feelings includes every possible variety of feeling, all the different kinds of sensation, all the different kinds of emotion, and all the different kinds of pleasure and pain which arise either in the one or in the other. We never have a feeling of any kind but as part of the succession.

But there are two kinds of sensation which are so peculiar, and from the formal element which they contain so important, as to require to be separately mentioned; the senses of sight and touch, including in the latter the sense of resistance, pressure, and effort. The sensations of these senses are part of the train of consciousness, and therefore are in time; but they also have extension, that is, occupy two at least of the so-called dimensions of space. There are certain sensations of both kinds which cannot be felt except in and with extension. The space element, superficial extension, is inseparable from these sensations. Imagine or try to imagine them without superficial extension, and you abolish the sensation at the moment you abolish the extension. Their inseparability from superficial extension stands the test of presentation. No one has ever *seen* a mathematical point; no one has ever *touched* one. On this account these sensations are said to be in the form of space, just as all sensations, these included, are in that of time.

Book II.
Ch. IV.
─ ─
§ 3.
Analysis of
minima of con-
sciousness.

The importance of these two senses, and especially that of touch in the large meaning here intended, is enormous.   Physiologically it is the
groundwork of all the rest; psychologically its
notices are the test of theirs. Of the two, touch is
the most irresistible in its evidence, sight the most
explicit and delicate. Touch is the text, sight is the
comment. Touch gives the meaning, sight interprets
it. Touch is force, sight subtilty.

But neither the minima of sight nor those of
touch involve more than superficial extension; they
do not involve solid extension, the so-called third
dimension of space. Solid space is not given in or by
isolated perceptions of sight or touch. It is built up
out of presentations and representations together, by
means of a reasoning process, and it involves the
perception of motion. Motion is to the perception of
solids what change is to perception generally. Motion may be defined as change in percepts of sight,
touch, or both.

It is a much discussed question, what part is
played by the two senses of sight and touch in giving
us our perception of solid bodies, and of space in
three dimensions; and whether either of the two
senses, without the other, is sufficient for that purpose. The question relates to the precise genesis and
history of our perception of matter (in the ordinary
sense of the term); and is of the highest interest.
For matter and motion are the two conceptions which .
together are the basis of all physical science. Everything is scientifically explained, when it has been
accounted for either as being itself a case of matter
in motion, or as depending upon one, by being referred to laws of material motion which are already

PRESENTATION AND REPRESENTATION. 267

Book II.
Ch. IV.

§ 3.
Analysis of
mimima of con-
sciousness.

known. There is no efficient known cause in existence, other than this. Of course this does not imply
that either matter or force is an ultimate entity.

But here I make my bow to the question. It is
necessary to indicate the place it holds in philosophy,
and to show at what point the conceptions of space
in three dimensions, and of solid bodies, conceptions
which in all probability are simultaneous, come into
human experience. We find ourselves already in a
world of solid bodies and solid space, on emerging
from infancy and beginning to be conscious of ourselves. We find ourselves in a world of sensations
already formed into coherent aggregates, "remote
objects" as I called them in a previous work.[1] How
was this finding brought about, how much was elementary, how much hereditary, how much acquired
by the individual? And in the case of the race,—
how much was elementary, how much acquired?
What moreover is the part contributed by the several
elements, that is, the two senses in question, and
their doubly elementary constitution? These questions are mainly psychological; certainly can only be
settled by psychological methods. Metaphysical analysis may be useful in preparation of the question,
but the solution is not hers. It is a question of
genesis, where physical and physiological considerations are supreme.

It will I know be objected that, seeing space in
three dimensions, or solid space, is not a *datum* but
a result, coeval with that of solid matter, the metaphysician has no right to speak of a space element in
all consciousness, but is misleading his readers unless
he restricts himself to speak of a superficial-extension-

[1] Time and Space, § 26.

BOOK II.
CH. IV.

§ 3.
Analysis of
inima of con-
sciousness.

element, or at any rate of an extension-element, in consciousness. But I hold that the metaphysician is perfectly justified in using the term space-element, in order to make clear to what the formal element in sight and touch sensation belongs, and of what it is finally, and not proximately merely, the rudiment. Proximately it is the rudiment only of superficial extension ; but finally it is the rudiment of space in the largest sense of the term.

Space is a perfectly general term, including under it metaphysical, or unfigured, as well as mathematical, or figured space. To renounce this use of space as a general term would be to admit that there is no space but mathematical, figured space; in other words, that an empirical *ens imaginarium* was an ultimate source of consciousness. But it is I think indisputable, that mathematical space is not such an ultimate source, but that it is a case of a larger and more elementary conception, namely, spatial extension at large, without any mention being made of dimensions, or directions of measuring or traversing it.

Metaphysical space is abstract capacity, Giordano Bruno's *res quædam nata repleri*. All that we positively know of it is adequately represented by what are called its three dimensions, which include all which are known to us. And therefore it may often happen that abstract capacity may be explained to mean space of three dimensions, this latter being taken as the *denotation* of the former, to give definiteness to our meaning. But it is not to be inferred that this abstract capacity is wholly and necessarily exhausted by the three dimensions. Space of three, or of *n*, dimensions is already mathematical space; for dimensions are only given by means of a distinc-

Book II.
Ch. IV.

§ 3.
Analysis of
minima of con-
sciousness.

tion introduced into abstract capacity, that is, by the addition of a difference of feeling.

All mathematical construction of space lies within or below this large conception of metaphysical space. Whether we adopt or reject what is called "Euclid's Postulate" is entirely indifferent to metaphysic. If, rejecting it, we construct a space bounded either by a spherical or by a pseudo-spherical surface, then we have the infinite extension of metaphysical space beyond it and containing it; beyond it on all sides in the case of spherical surfaces, and beyond it on some sides in the case of pseudo-spherical. All figures of space are finite. A closed figure is a finite figure; and at whatever part a figure is closed, at that part it is bounded, and has space beyond it.

Now the word extension alone does not convey this large general notion, it requires *spatial* to be added to it; or in brief to be spoken of as *space*. I maintain, then, that I am perfectly justified, inasmuch as the suggestion thereby made is true, in using the term space element for the formal element in visual and tactual perception.

6. Lastly I have to observe the relation between what I call the formal and the material elements in the succession of different feelings which we are examining. The *minima* in that succession are percepts, and these percepts are the ultimate empirical objects of metaphysic, and are analysable into the inseparable metaphysical elements, feeling and duration in all cases; feeling, duration, and superficial extension besides in some cases. These elements of time, space, and feeling, are the metaphysical ultimates of analysis. But the percepts as produced by shocks, stimuli, or changes in the relation between

Book II.
Ch. IV.
————
§ 3.
Analysis of
minima of con-
sciousness.

nerve-forces and stimuli, and in this way quantifiable and measureable, are the ultimates of psychology. The two features "produced by" and "measureable" show them to be the object of a science, psychology. Feeling, in all its kinds, and however named, e.g. a sensation or an emotion, means in metaphysic not the same thing as percept, but the inseparable material element in percepts.

But along with this analysis of percepts into their metaphysical elements, the foregoing examination has shown that there are elements of another kind in the minima of consciousness. There are in every minimum two sub-feelings, as I have called them, one on each side of the present instant, the time-point, which is the instant of change between them. These are elements of another kind from those which I call the metaphysical elements. They are properly *parts*, not elements. Very minute parts it is true, so minute as not to be perceptible alone, but only in conjunction. Still the difference between them and the whole they compose, and also between the parts themselves, is a difference of degree, not of kind. They are former and latter in time; and the imperceptibility of each part alone shows only that time is more minutely divisible in thought or imagination than it is in actual presentative perception. If the acuteness of our perceptive powers were increased, we should be able to perceive smaller minima, which would then have smaller parts or sub-feelings, than at present. We should then also perceive minuter differences of feeling than we do now, have a richer world of sensation and emotion than we have now, even without supposing any new sense or new emotion added to our list.

PRESENTATION AND REPRESENTATION.   271

Book II.
Ch. IV.

§ 3.
Analysis of
minima of con-
sciousness.

The difference between the time and space elements of percepts, on the one hand, and the feeling
element on the other, is not a difference of this sort;
it is a difference of kind. These two sorts of elements
are irreducible one to the other. Cut the train of
consciousness where you will, you find these two
sorts of elements, inseparable, irreducible to each
other, and simultaneously present; not as former and
latter, or escaping consciousness only from their
minuteness. They are the ultimate analysis of consciousness κατὰ μέρη and not, as the sub-feelings are,
κατὰ μέλη.

The question has been raised whether the time
and space elements are properly called formal, and
the feeling elements material; whether, even if these
names are retained, their application should not be
inverted. But I think that, even if Kant's usage were
not alone decisive on a point of this sort, there are
sufficient reasons for the names and their application as at present. The time and space elements,
duration and extension, are that which becomes order
and figure when distinguished by differences of feeling. They are order and figure in their rudiments,
are the condition of time and space *relations* between
feelings, are that from which form and relation come.
Feeling on the other hand is that which fills and
divides duration and extension, that which makes a
plenum of their *vacuum*.

This applies to consciousness generally, and therefore in metaphysic I think it most appropriate to
keep to the present usage. But when we come to
the sciences, mathematic and those which depend
upon it, then we come also to a specialised kind of
time and space; and here it may possibly be proper

Book II.
Ch. IV.

§ 3.
Analysis of
minima of con-
sciousness.

to change the application of the terms. In geometry we have to do with figured space, in calculation with discrete time. The figuration and the discreteness are derived from feeling, the material element in consciousness. And the object-matters which result, figured space and discrete time, are no longer metaphysical elements of consciousness, but empirical objects, *entia imaginaria*, the substrates of more specialised figures, numbers, and proportions. So also in the physical sciences, in which come forward additional determinations of the same substrates, by means of the presence of some kinds of feeling and the absence of others. Space and time, so determined, are physical objects. Thus Maimon says:[1] "With me, then, empirical objects given as outside one another are nothing else than space determined by empirical marks (*Merkmale*). Space is therefore the matter, the determinable (*bestimmbare*), and the marks of the empirical objects which occupy it are the form, or particular determination of the space." But Maimon adds, that since the critical philosophy has already fixed the meaning of the expressions, matter and form, he will use expressions of his own instead of them, namely, the *determinable* and its *determination*.

If we draw the distinction for which I contend between science and metaphysic, then I think we may retain Kant's usage for metaphysic, while adopting Maimon's for science. The time and space elements will still be the formal, feeling the material, element in consciousness generally, yet without attributing to either kind an *a priori* or transcendental character, as conditions existing previous to perception.

[1] Die Kategorien des Aristoteles : Propädeutik, p. 248.

PRESENTATION AND REPRESENTATION. 273

BOOK II.
CH. IV.·

§ 4.
Analysis of the
train of
percepts.

§ 4. The foregoing six heads complete the first part of my promised analysis, which has been an analysis of the succession of different feelings into its inseparable or strictly metaphysical elements. Another analysis has still to be given of the same succession, considered as a sequence of empirical percepts passing through consciousness, as a train of redintegration. I no longer take the minimum, the smallest fixable portion of the train, to see what elements it comprises; but I take the whole train as consisting of different empirical portions, and see what are its characteristics as a train. I still make abstraction, as before, of any specialty in the mode in which the train becomes distinguished into portions; that is to say, of any kind of difference other than the perceived differences in feeling already spoken of. We cannot help having a train of different feelings; I abstract from all volitional modification of the train. In other words, I take as object of analysis trains of spontaneous, not voluntary, redintegration.

This second analysis falls into two divisions, 1st, the nature of the nexus between an empirical present and an empirical past or future, which may be called the Problem of Memory; and 2nd, the nature of the order of sequence between empirical states of consciousness generally, which may be called the Problem of Association.

I. The Problem of Memory. The train of redintegration is a sequence of empirical differents. The percepts come in a series. But among them there are some which seem to be *repetitions*, to be the *same* that we have had before, either in a different train of redintegration or earlier in the same train. Now since the train is in perpetual change, move-

Book II.
Ch. IV.
——
§ 4.
Analysis of the
train of
percepts.

ment into the Past, two questions arise, 1st, How can there be such a thing as repetition? and 2nd, supposing there were such a thing, how can we ever know that it is a repetition, that the percept we have *now* is the same that we had *then?* This second question is the one urged by Mr. Ward, in the passage quoted above.[1]

There is, however, a question prior to both, which must be treated first. Until we know what *we mean* by repetition and sameness in redintegration, we cannot even enquire how they are brought about; and the analysis of what we mean by these terms will tell us what sort of evidence we are to expect for their possibility, which is the second question. And then will come the first question, how repetition and sameness are possible.

What we find, then, in redintegration is, to take Mr. Ward's instance, a present impression of having been cold a short time, say half an hour, ago. There is a present representation of cold, and a present representation of half an hour ago, that is, of the *surroundings* of a present representation of cold, which are not the surroundings of present presentations. The present representation of cold appears in *two* surroundings, those called *now*, and those called *half an hour ago;* or, in other words, there are *two* present representations of cold, present side by side in consciousness, each with its own surroundings. The two present representations of cold are indistinguishable in point of quality or kind; but they are in two different surroundings. That is what we mean by sameness in kind, and difference in number. It is also a case of *repetition*, for the two sur-

[1] At page 251-5.

Book II.
Ch. IV.

§ 4.
Analysis of the
train of
percepts.

roundings are not only different, but different as former and latter in time. It is clear, too, that there might be sameness in number and difference in kind, if the difference lay in the quality without any difference in the surroundings. But this would not be a case of redintegration of empiricals; the sameness in the surroundings restricting us to a single portion of the stream of consciousness.

This, then, being what we mean by sameness and by repetition, namely, feeling indistinguishable, formal setting different,—it is clear what sort of evidence is required to show that repetition has really taken place in redintegration. The feeling must, in the moment of memory, be capable of being *twice* brought into consciousness, once in each of its two surroundings. The moment of memory is a moment of superposition of one instance of the feeling on the other.

The continuity and discreteness of the formal element in consciousness at once connect and hold apart the various empirical portions of its stream. Together they are a condition both of sameness and difference in empiricals. Without continuity in the formal element, the empirical states of consciousness, even supposing them to exist, could not form part of a single whole; without its discreteness, they would fall together into a chaotic mass. They are a condition of having redintegration at all. And therefore I said above[1] that future analysis would show, that sensation and consciousness suffice to explain the validity of memory, if only they are examined by the reflective method, and not arbitrarily divided into separate empirical states.

[1] At page 255.

Book II.
Ch. IV.
───
§ 4.
Analysis of the
train of
percepts.

We are now prepared for the first question, What are the conditions of repetitions taking place in redintegration? Evidently, for one thing, a continuation of the conditions for presentations taking place. The nerve process which has subserved a presentation can be set on foot again by a new stimulus, which may come either from another part of the nervous organism, or from the world outside the organism; and in either case we have two nerve processes superposed, exactly similar in point of the quality of feeling which they subserve. The two nerve processes subserve the two representations, which are the content of the perception of sameness.

This may be illustrated by the explanation sometimes given of what is called illusory memory. This phenomenon is where a man, while having actual presentation of a scene or an event, in which he takes part, has the strong impression of having seen or done the very same things before, and yet knows, by reasoning, that he never has seen them or done them. This strong impression is sometimes explained as a case of doubleness in the organs subserving perception, a double set of images being brought into consciousness, and not perfectly simultaneously, whereby they are prevented from coalescing into a single object, as the images in the stereoscope do.

Now wherein lies the illusion in the case of illusory memory? In this that, while the images are *two*, the surroundings are one and the same, at least are not perceptibly different. The Subject of illusory memory, sensible of two images, refers them to different times, or to different times and places, just as he would do with phenomena of ordinary memory. He refers one of them to the past, the other to the

Book II.
Ch. IV.

§ 4.
Analysis of the
train of
percepts.

present, instead of referring both to the present, as he would do if he was aware of the unequal action of his nervous organism in the particular case.

It may possibly be imagined that since reasoning is involved, at least apparently, in cases of illusory memory, and certainly in many comparatively simple cases of ordinary memory and perception, as for instance where we either see, or remember seeing, a set of feelings and the class to which they belong together, as a ship, or a house,—it may be imagined that therefore every case of memory, which is equivalent to saying every case of perception, involves reasoning. Unconscious reasoning, or unconscious inference, is the way it is sometimes expressed. And this theory is own brother to, if not identical with, the theory of a transcendental reasoning faculty, to perform the unconscious reasonings.

Now no reasoning process can, as such, be unconscious. True, a conscious reasoning process may become habitual and be performed unconsciously; but then it ceases to be a reasoning process. It becomes a process of complex perception. Unconscious reasoning is a contradiction in terms. But it may be said, This is all we contend for. Has not conscious reasoning entered into all perception originally, and left habits behind it, which then, being thenceforth unconsciously performed, are properly called perception and not reasoning?

This is undoubtedly the fact in a very large number of cases. But it cannot be extended to all. It is no pre-requisite of perception or of memory altogether. Just as the notions of time and space presuppose time and space elements in perceptions, so a reasoning process presupposes a perceptive pro-

Book II.
Ch. IV.
——
§ 4.
Analysis of the
train of
percepts.

cess to begin with. or else we could give no account at all of the reasoning process. Observe, since I am not tracing the *history* of conscious development, but analysing it, I must not be understood to assert that volition is not found among the earliest states of consciousness; but my meaning is that, whenever it appears, it has before it a previous state of spontaneous redintegration, out of which it springs by a single modification which will be examined in the following Chapter. And so we find that in memory there are elementary trains of redintegration, which do not involve reasoning at all. And this involves the priority of spontaneous to voluntary redintegration, which is in other words the priority of perception to reasoning.

Let us examine the case as one of memory. In remembering that I was cold half an hour ago, I do not *argue* that one of the representations of cold must be in the past, because it does not fit into present surroundings; this would imply that I had an *a priori* knowledge of past and present time; but I *perceive* the two representations of cold at once, in their different surroundings, as portions of a connected train, the intermediate members of which can be brought back into memory also. It is simply a lengthening of a minimum of presentation, such as was analysed above.[1] The minimum of presentation was there said to contain the rudiments of memory. It can be no case of reasoning, for where are the premisses? And if it is said to be a case of induction, which is a process of reasoning. I answer that induction is not a process of reasoning alone, but is a *method*, consisting partly of perception and partly of reasoning; and that the elementary cases of memory are the per-

[1] In § 3.

Book II.
Ch. IV.

§ 4.
Analysis of the
train of
percepts.

ceived facts which are the basis of the method of induction, and the premisses of the reasoning process which completes it. It is of the essence of induction to lay percepts at the basis of reasoning, and not to use reasoning, without percepts, to form a basis for itself.

In conclusion I will observe, that the foregoing account of memory is a combination of a metaphysical analysis of its phenomena with the (very imperfect) assignment of its physiological conditions; and that the chief burden of explanation is borne by the metaphysical analysis. The physiological part of the account, taken alone, however perfect our knowledge might become, would be quite insufficient as an explanation, because it assumes the meaning of sameness and repetition to be known, and assumes farther that nerve matter and nerve processes exist as permanent realities, which they cannot do except on the supposition that sameness and repetition are valid conceptions; for they are portions of the visible and tangible world, which cannot be thought or spoken of without sameness and repetition being thought as part of them. The physiological part of the account, then, is subordinate, being restricted to the task of assigning the genesis and history of the phenomena in question, but incapable alone of exhibiting their nature and making manifest their logical consistency.

II. The Problem of Association. This question relates to the general features or nature of the order in which the empirical portions of a redintegration tend to follow each other, when left undisturbed by volition or conscious purpose. In a question of this large kind, the imperfection of the record makes the problem especially obscure and difficult. I mean by

Book II.
Ch. IV.
———
§ 4.
Analysis of the
train of
percepts.

the imperfection of the record the fact, that many nerve processes take place without being attended by consciousness at the time, the threshold of consciousness not being passed, though doubtless they are not without their effect in modifying future states of consciousness when they arise, by entering into and modifying those nerve processes which are attended by consciousness. We have every reason to believe that, while every state of consciousness depends at the time of its arising upon nerve process, it is not the case that every or nearly every nerve process is attended at the time with consciousness.

This problem also falls into two parts, first, the analysis of the order obtaining in redintegrations as sequences of conscious states; and second, the particular condition in nerve processes which determines that order to be what it is.

The first part of the question falls again into branches; the first of which concerns the *direction* of the sequence, in respect to first and last. Is the direction of the sequence in representative redintegration necessarily the same as in presentation, or may it vary; may not the order of sequence in presentation be reversed in representation? Experience tells us at once that it may be reversed. We are not restricted in memory, and still less in imagination, to repeat the phenomena in that order of sequence in which they occurred in presentation.

Let us put the matter into schematic form, by throwing it into a sort of general mental diagram. Suppose I have a series of presentations which I will call Red, Green, Yellow, Blue, in that order. Then suppose that something which I *hear* recalls into

BOOK II.
CH. IV.

§ 4.
Analysis of the
train of
percepts.

consciousness Blue; and this must be by a stimulus from within the brain; then the representation Blue may call up Yellow, Green, Red, the stimulation proceeding from Blue. The order of representation reverses the order of presentation.

Just the same will happen if Blue is called back into consciousness by a new presentation. It will recall Yellow, Green, Red, in that order, as in the previous case.

But if Red is the starting point of the representative series, whether called back from within or from without, the sequence of representation will be in the same direction as the sequence of presentation.

The direction of the sequence in representative redintegration is therefore in great measure independent of the direction of the sequence in presentation. What is first in the one may be last in the other. We may picture the redintegrations flashing as it were from point to point along any of the lines of nerve communication in the brain, or in the whole aggregation of organs which compose the nerve organism. The order of revivification may be quite different from the order of original vivification. But the standard of comparison in respect to first and last is always the order of presentation. When we say that Red, for instance, is really first, and Blue last, we mean that they were so in presentation. There is no other standard possible.

The domination exercised by the direction of presentation over that of representation is shown by some phenomena of purely representative imagination, as in dreaming, where we always construct the story as if it were a presentative history, notwithstanding that the order in which the representative

Book II.
Ch. IV.
___
§ 4.
Analysis of the
train of
percepts.

images really take place in the brain is different or reversed. In the *Theory of Practice*[1] I quoted a striking dream related by M. Maury, in which this unconscious reversal of the representative into the presentative order was made evident.

There is no *absolute* first and last in time, just as there is no *absolute* up and down in space. Still the determinations of first and last, of up and down, are not given in the first instance by reasoning, but by spontaneous redintegration, and afterwards interpreted by reasoning. We have them *in their first intention* in spontaneous redintegration. It is the order of presentation that we call the *real* order, when we have contrasted it, by reasoning, with the order of representation. There is nothing more ultimate than the order of presentative perception. And this order is the final criterion of first and last in perception generally, just as the presentations of touch are the final criterion of up and down, front and back, right and left, in spatial perception. Notwithstanding the inverted incidence of visible objects on the retina, we *see* them upright because we identify the spot touched with the spot seen. So in dreaming, we *experience* the incidents in historical order, notwithstanding that they may have occurred to consciousness in *any* order, and the actual order may in some cases be pointed out.

Turning next to the second branch of the question, we find that it relates to the actual order in which spontaneous representative redintegration takes place, abstracting from presentations, and from the different direction of the sequences in respect to first and last. It is the " Laws of Association of

[1] Vol. I. p. 115. note.

Book II.
Ch. IV.

§ 4.
Analysis of the
train of
percepts.

Ideas" that we have to investigate. The old laws of association as given by Hume, namely, Contiguity in time and place, Similarity and Contrast, and Cause and Effect, are manifestly insufficient. What we want is, not a conflict of laws, but a single law, under which more special laws may be grouped as cases. And now let us look at the facts.

The first law is plainly nothing else than this, that the redintegration itself is a sequence at once continuous and discrete, continuous in one respect, discrete in another. This law depends upon the analysis of minima, given above. But the next question is, What determines this sequence to take place in one order rather than in another; what is the second law of redintegration?

We are perfectly justified in putting aside and abstracting from the interference of fresh presentations of sense, in considering only the order which representative redintegration tends to take when left to itself. It is indispensable to do this, because otherwise we should not be examining spontaneous but enforced redintegration, not the course of consciousness and its organism left to themselves, but the course of consciousness and its organism and the world external to the organism together. It is, then, within purely representative redintegration that the second law must be sought. And here I shall content myself with briefly stating the result to which I came in my two former works;[1] to reason it out over again would lead me into a discussion out of all proportion to the present work. As before, I put forward the following as an hypothesis only, as a pre-

[1] Time and Space, §§ 28. 29. 30. The Theory of Practice, Vol. 1. § 53.

Book II.
Ch. IV.

§ 4.
Analysis of the
train of
percepts.

liminary analysis still requiring completion and verification.

It seems to me, then, that the order of sequence
in trains of spontaneous representative redintegration
may be conceived as a resultant of two sets of
features in the feelings which compose the redintegration, one of which sets is subdivisible into two
again.

The first or double set consists of feelings as
marked by, 1st vividness, 2nd frequency. First, a
great degree of vividness or intensity in a feeling
makes it more likely to recur again; and those feelings which are most vivid tend to recur most frequently. Secondly, the oftener a feeling has been
experienced in the past, the more likely it is to be
experienced in the future; past frequency, counting
from any moment, is a mark that greater frequency
may be expected after that moment. This is what
we call Habit. And we may include habit among
the fundamental features of spontaneous representation without violating logic, because we are not considering the origin of the order of redintegration, but
merely its analysis. If spontaneous representation
were a self-contained and independent phenomenon,
not influenced by presentations, then the circumstance of frequency of recurrence in its feelings
would not be an ultimate fact in it, but would
have to be traced to some further source. But as
it is, spontaneous representation has one source at
any rate in presentations, from which we are only
artificially isolating it; and the connection of spontaneous representation with presentation is not now
our object. Both vividness and frequency have their
ultimate source in presentations; but we are now

considering them only as occurring in representative sequences.

The second set consists of feelings of some specific pleasure or interest. And this set supplies the counterbalancing tendency to the other set with its two branches, vividness and frequency. While vivid feelings and habitual feelings tend to perfect uniformity in redintegrations, feelings of pleasure and interest tend to variety, to deflect the habitual current of percepts.

Between them, these features determine the order of the redintegration. But of course they mix in the most various degrees, and are combined with other feelings of the most various kinds. Some feelings may be habitual without being vivid or pleasureable; others may be pleasureable but not vivid; and so on in endless variety. If a feeling is at once extremely vivid, habitual, and pleasureable, its chance of recurrence is enormous, and as a motive it is almost irresistible.

So much as to the subjective analysis of spontaneous redintegration. Now for the second question concerning it, namely, the physiological conditions which would enable us to give its psychological analysis. In the first place, the nerve processes supporting representations must be the continuations of those which support presentations. This has been already remarked, under the problem of memory. The necessity is the same in both cases.

But when we come to the more special modes of action which characterise the nerve or brain processes, underlying redintegrations which are supposed isolated from new impressions of sense presentation, the state of physiological knowledge gives us but

Book II.
Ch. IV.
§ 4.
Analysis of the
train of
percepts.

little if any light. Here, in fact, we know the subjective phenomena better than their objective conditions. And here, accordingly, metaphysical analysis may fairly be taken as a guide to psychological, by directing physiological research.

What I mean is that, supposing the analysis of spontaneous redintegration, given above as hypothetical, to be correct, it tells the psychologist what questions to put to the facts of physiology which he examines. He must look for nerve processes which are the conditions, severally, of the phenomena of vividness, of habit, and of specific pleasure or interest. If it is permitted to anticipate the results of such an examination, there is one result at any rate which I am not alone in thinking highly probable; it is that the nerve processes which underlie the feelings of pleasure or interest are identical with those which are the main or positive factors in the healthy growth and development of the organism. The mere sense of *bien-être* implies and depends upon a healthy organism; and, if the present hypothesis is correct, we should but have discovered the earliest and most rudimentary manifestation, within consciousness, of the great law which runs through the organic world, moral and physical, that a sound mind is inseparably dependent on a sound body.

For when we analyse redintegration, which is consciousness in forward moving action, it is not only memory and imagination of which we are tracing the rudiments. The laws of association, being laws of consciousness in *action*, give us also the rudiments of the Will. *Spontaneously*, if the present hypothesis be correct, the conscious healthy organism seeks pleasure and avoids pain, and this action is identical with the action of self-preservation and healthy growth, such as might take place without consciousness, and such as does take place in the vegetable world. The *spontaneous* play of feelings in redintegration, the spontaneous interaction between vivid and habitual feelings, on the one side, and pleasureable feelings on the other, is a state prior to volition, containing the rudiments out of which it springs by a certain superinduced modification. Volition is conscious choice between those same feelings which compose the current of spontaneous redintegration; is that same action modified by a reaction of the organism itself; how modified and to what result, we shall partly see in the following Chapter.

Book II.
Ch. IV.

§ 1.
Analysis of the
train of
percepts.

CHAPTER V.

PERCEPT AND CONCEPT.

Book II.
Ch. V.
— —
§ 1.
Analysis of the
conceptual
process.

§ 1. THE step which we are now to take is one which carries us at once into the very heart of metaphysical controversy. We have again to apply reflection, as in the preceding Chapter, and again to make primary consciousness, its states and processes, the main object of examination. At the same time we shall have to take into account those of direct, and still more those of reflective consciousness, because the act or moment of conception in primary consciousness is the moment in which reflection itself arises, and it will be necessary to contrast the two cases, of conception with and conception without, or previous to, reflection.

In the last Chapter abstraction was made from the means by which the succession of different feelings was broken up into portions. It was taken as a train of percepts, presentations and representations together; and every part of the train, whether taken to consist of pure presentations, or pure representations, or of a mixture of both, was called a percept. The name for such trains of percepts, consisting as they must do to a very great extent of representa-

Book II.
Ch. V.
———
§ 1.
Analysis of the
conceptual
process.

tions, is *redintegration*. And hitherto we have been occupied solely with that kind of it which, to distinguish it from the volitional kind which is next to be examined, may be properly called spontaneous redintegration. Conception is, it will be seen, a case of voluntary redintegration.

Observe a feature of the method which is common both to the foregoing and to the present enquiry. It is the combination of the statical and the dynamical way of treatment. In the foregoing analysis of the succession of different feelings, we began by considering the minima, and took these portions statically; but we found that they all and each contained a succession, a process, that is, might be treated dynamically. In the present case, we begin with the notion of process, and treat the train of percepts as a redintegration. We shall find that it will break itself up into portions, not the same in kind as those with which it begins, but into statical portions of another kind. We never get beyond these two ways of treating phenomena, the static way and the dynamic. Together they are exhaustive. It is so in the sciences, for instance, in mechanic. Statics treat the same mechanical phenomena as dynamics, but making abstraction of *time*; and that is the difference between the two branches. The reason why statics and dynamics are felt to be an exhaustive treatment of mechanical phenomena is, that consciousness itself, in its utmost simplicity, moves in the same rhythm.

The first question which occurs in regard to redintegration has been already answered in the last Chapter, namely, what distinguishes one state of consciousness in the redintegration from another, or where does one state end and another begin ? The

Book II.
Ch. V.
———
§ 1.
Analysis of the
conceptual
process.

answer was, that wherever there is a difference of *feeling* there is a difference of state. Until a difference of feeling occurs, a state continues one and uninterrupted. The chain of consciousness in redintegration is a sequence of Differents.

The next question, to which we now proceed, is this : What breaks up this continuous stream of differents into separate portions, some of greater, some of less, complexity, and forms out of it that congeries of objects which we call the world ? There are two alternatives usually proposed as answers to this question, or rather as directions in which to look for the answer. The first is, that " we," the *ego*, must have a power working in a method by which the transformation is effected. The second is, that somewhere in the objects themselves there exists a force, and a law of that force, which is prior both to the continuous stream and to the objects as we conceive them, a force and a law inaccessible to us, which produces both the stream and the objects.

Without dwelling upon what is unsatisfactory in each of these modes of finding an answer to the question, I confine myself to remarking that both answers are answers to the question of *genesis*, and not to the question of *nature*. The answer which I shall give is an answer solely to the latter question. When I ask, " What breaks up the continuous stream," I do not mean, " What power or force breaks it up;" I mean, What feature *in the continuous stream itself* is the invariable antecedent or condition of its being broken up into separate portions or objects ?

Taking the question in this sense, we shall find that there is such an invariable antecedent comprised in the continuous stream of redintegration. We find

Book II.
Ch. V.

§ 1.
Analysis of the
conceptual
process.

that we *attend* to some states and disregard others in the chain, and that this modifies the chain. I use the words *we* and *attend* denotatively, and in no other way; they do not imply an ego or an agent; they are the only means I have of pointing out to my readers the particular phenomenon I intend. The fact or moment of attention is that which distinguishes voluntary from spontaneous redintegration. For attention, when guided by a purpose, is an exercise of volition. "Preliminary to every act of thought," says Mansel,[1] "is an act of Will, *attention*, in which the mind contemplates exclusively a certain number of the attributes given in an intuition to the neglect of the rest. By thought these attributes are regarded in their relation to objects." And in a footnote he speaks of "the voluntary element of *attention*" as "an element neglected by the Kantian as well as by the sensational school, and only fully appreciated since the reaction against the latter, commencing with the lectures of Laromiguière."

Let us see, then, in the first place how this moment of attention is introduced into the chain of spontaneous redintegration. We are not conscious of the minima, analysed in the last Chapter, in the first instance. The train does not come to us broken up into separate minima. That is an artificial division introduced by the analyst. The train comes to us in masses; mostly containing many different feelings; it is therefore not mere difference of feeling that breaks the continuity of the train, or suffices to constitute a separate portion of it. The greater differences only can do this, the more striking contrasts of feeling. It is precisely these

---

[1] *Letters, Lectures, and Reviews*, p. 48. Recent extensions of Formal Logic.

BOOK II.
Ch. V.
—
§ 1.
Analysis of the
conceptual
process.

to which we attend in the first instance, that is, before we have any pre-conception as a standing element in the train, and acting to modify its course by giving an intensity to particular feelings, which they would not have by themselves. A flash of light, a loud sound, and so on, intervening in the redintegration, arrest the attention, as it is called; we pause to say, what is that? We dwell upon it and keep it in representation, long after it has died out of presentation, that is, has ceased to be actually seen or heard.

The reaction, expressed in the *what is that?* is the precise moment of attention. Not the tension or sense of effort, or nerve-sensation, which accompanies the flash or the sound; but the reaction upon it. But this reaction is conditioned and determined by a mere difference of degree, a difference of greater and less in feeling. The reaction, the attention itself, is the beginning of a difference of *kind* between those states which precede and those which follow it.

Consciousness itself appears to involve some reaction on our part, on the part of the organism. Otherwise we should be having feelings which nobody felt. To feel is to react. Pure passivity is as impossible a notion as pure activity. But it does not follow that we are conscious of the reaction as such. This reaction may be accompanied by a feeling which as a distinct feeling is faint at first, then rises and strengthens till it reaches the threshold of distinct consciousness; a feeling accompanying every other more special feeling, and taking a colour from it. When it rises to the threshold of consciousness, it is nerve-feeling, sense of tension or effort, or in German *Innervationsgefühl*.

But when this nerve-sense rises to a particular

Book II.
Ch. V.
§ 1.
Analysis of the
conceptual
process.

degree of intensity, or stands in a particular contrast with other feelings, it is found to undergo a modification; from being a nerve-sensation it becomes a sensation of attention; it is as if the reaction of the nerve or nerve organism was felt *as a reaction.* And no doubt there is, underlying it as its physiological condition, a different and farther reaching nerve-action than when we had the simple nerve-sensation alone in consciousness. There no doubt is a new action set up in, and beginning from, the central part of the nerve-organ affected; and not simply a reaction of the nerve on its stimulus, which we must conceive as requisite to sensation alone. There is, in short, one threshold of consciousness, and another of reactive consciousness, marked by attention. And this second threshold is a stage in the rising intensity of the particular kind of sensation known as nerve-sense or tension; a moment which is at once a heightening or intensifying of the feeling, in which it occurs, and a beginning of a process, new in point of kind, the process of reasoning, since it is a moment which supervenes upon and mixes with a succession of different feelings.

Carrying on the analysis of voluntary redintegration still farther, we observe two main and generic classes of purposes, the pleasure of feeling and the pleasure of knowing, which guide the attention, by determining its kind or direction. In saying *What is that*, at a flash or sound, we have a case of the pleasure of knowing. Dwelling on a pleasant sensation, as in tasting flavours, is an instance of the pleasure of feeling. Vividness in a feeling compels, pleasureableness attracts, the reaction of attention. All volition consists in this sort of reaction. What

Book II.
Ch. V.
———
§ i.
Analysis of the
conceptual
process.

are usually set down as a third class of volitions, volitions to act, constitute really the general class to which the other two. volition to feel or enjoy, and volition to know, are subordinate classes. There are two kinds of volition, and all volition is action. Volition to feel is practical, volition to know theoretical, intelligence. To attend in order to know is the kind of voluntary act which distinguishes *Reasoning*, as a mode of voluntary redintegration, from spontaneous redintegration the common source of both practical and theoretical intelligence.

And now let us see what modification the moment of attention introduces into the chain of spontaneous redintegration. It gives us the distinction between percept and concept as determinations of the chain; it gives us the chain no longer merely distinguished into portions by difference of feeling, but distinguished into portions, not the same as might have arisen without it, each of which has a double character, an actual character and an expectant character, actual as it is perceived in and by itself, expectant as it waits for the completion of its relationship to other portions. In its actual character every portion of the chain is a Percept; in its expectant character it is a Concept.

Observe, the fixing of attention on a single portion effects a change in the character of the whole chain. The fact is not that one portion becomes a concept, while the rest remain percepts; nor yet that the chain now consists of concepts that are not percepts, instead of percepts that are not concepts; but every portion is now both percept and concept, according as it is taken in relation to other portions or out of such relation. As I have elsewhere expressed it, objects

Book II.
Ch. V.
—
§ 1.
Analysis of the
conceptual
process.

considered in their relation to consciousness alone are percepts, while objects considered in a certain kind of relation to other objects of consciousness are concepts. This distinction is perfectly general, in the sense that it is applicable to any portion of the chain of redintegration, however large or however small. Reduce the portion to a minimum, you have a percept if you take it in relation to consciousness alone; and a concept if you take it in relation to other objects in the redintegration. Enlarge it to a maximum, the same holds good, for all maxima are relative, none absolute and final.

It is, then, by attention modifying spontaneous redintegration that the distinction between percept and concept arises. Let us now consider more closely how attention operates to that effect. Spontaneous redintegration gives us a chain of percepts or differents, percept following percept. Here it becomes necessary for us to apply the distinction between the two modes of primary and direct consciousness, as drawn in Chapter II., because the course followed by conception is very different in the two, and the terms in which it has usually been described relate only to its operation in the latter, the supposition being made that "objects" are already given to us before conceptual redintegration begins. But the fact of the case is, that "objects" are first formed by means of conceptual redintegration, in its first mode or stage of primary consciousness. Experience does not give us "objects" ready made, any more than it gives us numbers ready counted, spaces ready measured, or matter ready weighed. We have to do all this for ourselves; and the early period of an infant's life, his period of primary consciousness, is occupied in

Book II.
Ch. V.
——
§ 1.
Analysis of the
conceptual
process.

the formation, subjectively, of his world of "objects," which he does by means of attention modifying his trains of spontaneous redintegration, and testing them by presentations.[1] I take therefore in the first place the operation of attention on redintegration in primary consciousness.

The fundamental law of all reasoning considered as an action is the Law of Parcimony, because it is the practical law of all voluntary effort to do the most we can with the least effort we can. The law of parcimony is the parallel, in consciousness, of the law that movement takes place in the line of least resistance in mechanic. When reasoning is the operation in question, this practical law becomes specialised, namely, to group as large a content as possible under the fewest heads possible, for *Frustra fit per plura quod fieri potest per pauciora*, which is the law of parcimony. This explains why, supposing we have the power to do so, (and it will presently be seen in what this power consists), we actually do always group similars together, and dissociate them from dissimilars.

So that, as Professor Bain truly says, "The primary attributes of Intellect are (1) Consciousness of *Difference*, (2) Consciousness of *Agreement*, and (3) *Retentiveness*."[2] And when he adds, "Every properly intellectual function involves one or more of these

---

[1] It was my friend, the late James Hinton, who first pointed out to me, that infants pass the earliest years of their life in learning to use their senses and forming their world of objects, that it is a period of great intellectual activity with them, an activity which we may often obstruct by interfering too much with our namings and explanations of things. And the importance of the remark struck me as fully warranting the stress he laid upon it.

[2] Mental and Moral Science, Book II. p. 82. 2nd edit. 1868.

attributes and nothing else," all that is wanting here is to draw the statement closer, dropping the "one or more ;" for all the three attributes are involved in every intellectual operation. Every such operation may be described, in consequence of the analysis in the foregoing Chapter, as a consciousness of agreement and difference in a redintegration.

In the next place, let us suppose that some member of the chain of redintegration, some perceived sensation, attracts our interest and forces us to dwell upon it. We can dwell upon it and go on with the redintegration at the same time, that is, combine with it a new redintegration, by two expedients only, either we reject everything that is *dissimilar* from it, retaining everything that is *similar*, or else we reject everything that does not stand in immediate connection, of time or place, with it, retaining everything that does stand in such connection. The first expedient gives us, as the result of the redintegrations formed by it, *general objects*, or concepts, *e.g.* red as standing for and containing all the red sensations which we have perceived; the second expedient gives us matters of fact, *e.g.* that a red sensation has stood in immediate connection with another sensation, say a hardness, (as in the case of the infant's coral).

These two methods combined, and repeated in various combinations, every day and all day long, in primary consciousness, give us at last our world of things or objects. There is a double source of the total perception of the world of objects, the two expedients which give us, one general objects, terms, or concepts, the other matters of fact. *Red* and *Hard* and so on are general objects; the combination of *this* red with *this* hard, (by which I do not mean this

Book II.
Ch. V.
—
§ 1.
Analysis of the
conceptual
process.

red *thing*, this hard *thing*, but this particular case of redness, of hardness), and so on, are matters of fact, things, existent objects. Observe the difference between the two sources. Both are in accordance with the law of parcimony, but only the first, the expedient which gives us general objects, is in consequence of it. We should be forced to combine red with hard, whether there was a law of parcimony or not; the combination is a matter of fact, which we have no control over. But to group this red with that red is, though equally a necessity, yet a necessity solely of consciousness itself, a necessity of doing in the easiest way what we want to do. It is a case of causality by a final cause, a purpose, a wish, or a need.

I now come to the case of direct consciousness, in which objects are supposed to be already formed, and we are in presence of a world of objects such as grown-up people find it. The method of reasoning which we employ upon this world is essentially the same as before. Only, instead of taking a sensation as the thing fixed by attention in redintegration, we take an "object" of which we already know a good deal. The same law of parcimony governs our procedure, the same two expedients offer themselves. Only, as our starting point is no longer simple but complex, an object not a sensation, the course of reasoning is modified correspondingly. We can hold the object before us and yet continue to redintegrate in three ways. First, we can redintegrate all the modes of the general qualities which constitute it; we can analyse the object into its qualities. Secondly, we can redintegrate it with similar objects. Thirdly, we can redintegrate it with its conditions or its conditionates, that is, its coexistents, its sequents, its

Book II.
Ch. V.

§ 1.
Analysis of the
conceptual
process.

antecedents. Carry this third mode of reasoning out, and you have the various physical sciences.

Observe here again the double source of the reasoning. The first method is partly a consequence of the law of parcimony, inasmuch as the qualities have been grouped together, in the first instance, only by that law; but as to their grouping together in the object examined, that is a matter of fact independent of that law. The second method is purely a consequence of the law of parcimony. The third is purely a discovery of matters of fact.

Let us now go once more over the same ground, and examine these same two cases, primary and direct, of voluntary redintegration, but in a somewhat different respect. Instead of taking single members of the chain of differents to become the expectants, let us suppose that two or more such differents are taken; that is, that a change or passage from one to another, that is to say a *relation*, is taken as the expectant unit. The object is different from what it was in the first way of taking it, but the way of dealing with it is just the same as before. In primary consciousness we are dealing with a change or sequence of feelings, in direct consciousness with a change or sequence of "things." In the first case, we have again the same *two* ways of regarding them, either to classify such sequence with similar sequences, or to combine it with the sequences, or with the feelings, which are connected with it in time or place. In fact this is one of the methods which contribute to the formation of the notion of "things," and together

Book II.
Ch. V.
——
§ 1.
nalysis of the
conceptual
process.

with the first method is the antecedent, in our mental history, of the mode of direct consciousness, or consciousness of objects or things as existent.

In the second case, relations of direct consciousness, we have a sequence or change of objects or things before us. The same *three* methods are open to us here. Either we analyse such a change into its component changes; or we classify the kinds of change in objects, the modes of their motion; or we combine the change with its antecedent, coexistent, and consequent changes.

The two ways of taking the redintegrations are respectively statical and dynamical. One gives us the kinds of feelings and the kinds of things; the other gives us the kinds of changes and the kinds of motion or force, for what is force but the relative motions (or tendencies to motion) of masses or things?

We have now before us the analysis of conception, in its origin from perception, in both the primary and the direct modes of consciousness. It remains to draw some conclusions respecting both, which the analysis warrants.

In the first place we see that the whole redintegration is made to change its character by the introduction of attention. Notwithstanding that we do not and cannot change its content at our will, in the sense that we can choose what content it shall offer us, yet since we can and do reject what it offers, until it offers what suits our purpose, and this offering is our power of modification spoken of above,[1] the whole

[1] p. 296.

PERCEPT AND CONCEPT. 301

Book II.
Ch. V.
___
§ 1.
Analysis of the
conceptual
process.

chain formed in this way is a voluntary remoulding of the whole chain of spontaneous redintegration, the whole chain of mere differents as we had them in perceptual redintegration. We have, so to speak, cut up the chain into lengths and piled similar pieces together. We have done the same with the chain considered as a series not of links but of changes between links. We have got the chain in pieces which are now connected together, not as they are given to us in perception, but in an order of thought, the connection of which consists in similarity and graduated differentiation, determined ultimately by the facility with which we can hold a large content of consciousness together with little effort, determined by the law of parcimony. This new nexus between the links of the chain of redintegration is the conceptual order, the order of thought, as opposed to the perceptual order, which is the same (in kind of nexus) as that of nature, of which it is an imperfect and partial reproduction. But the content of the two orders is the same, there is no content in the order of thought which is not a derivative of something or other in the order of perception.

In the next place I observe that this analysis shows us, that the elements of Time and Space are not got rid of in the conceptual order, but reappear in it. A concept is a provisional image, expectant of its detail or of its accompaniments; if the image is of a visible and tangible object, or of something which takes place in one, then that image has parts of extension, just as much as the visible and tangible object of which it is a representation. If an event requires time for its accomplishment, then the provisional image, or concept, of that event includes time;

Book II.
Ch. V.
———
§ 1.
Analysis of the
conceptual
process.

and that not merely in the sense that our brain requires time to think it, but that the image is an image of an event objectively in time. It is very necessary to insist upon this presence of time and space elements in the conceptual order as well as in the perceptual, because, whether from a habit of understanding by Time and Space only mathematical time and space, time marked out into lengths and space into figures,[1] disregarding time and space as elements of consciousness, that is, in their metaphysical acceptation,—or from whatever other reason, —nothing is more common than to hear of things which are said to be out of Space and Time, have nothing to do with them, but are beyond them altogether; such, for instance, as Eternity, the Moral Law, Hegel's True Infinite and Absolute Mind, or, as we saw in a previous Chapter,[2] "the presence of consciousness to itself." Of all loose talk in philosophy, none seems to me looser or more unjustifiable than this.

Thirdly, I remark that, while for the purposes of our knowledge we transform the perceptual into the conceptual order, the perceptual order does not cease to exist, but is the order into which we must re-transform the conceptual, whenever we verify our reasonings. If the object-matter is physical, we verify by observation or experiment, that is, by sense presentations. If it is mental, we verify by repeating

---

[1] As an instance of what I mean, take Mr. Spencer's expressions: "that consolidated abstract of relations of co-existence which we know as Space, and that consolidated abstract of relations of sequence which we know as Time." Principles of Psychology, Vol. I. p. 183. 2nd edit.

[2] Above, p. 159.

Book II.
Ch. V.
———
§ 1.
Analysis of the
conceptual
process.

the original redintegrations. In both cases we verify by a reference to the perceptual order. We bring the conceptions we have formed to the test, by comparing them with the perceptions they spring from, and this we do in Reflection upon them.

The next remark relates to one of the "two expedients" which are common to both primary and direct conception, to that one by which, holding fast the first concept by attention, we redintegrate it with its coexistents, sequents, or antecedents, in time and place. Of this method it is to be remarked, that it is a return to the perceptual order from the conceptual, and through the moment of conception. The redintegrations are no longer spontaneous but voluntary, and yet the objects which they give us, the successive members of the redintegration, are percepts and occur in the order of perception. They come before us as matters of fact, occurring in the same order as things and occurrences in presentative experience.

By adopting this, the second of the two expedients, what we have done is, that we have in one respect *reversed* the choice made of the conceptual order by the act of attention; we have first adopted the conceptual order by an act of attention, and then chosen to have objects as percepts and in perceptual order, notwithstanding that act. Of their *double* character, as percepts and as concepts, we have chosen the perceptual to have them in.

The voluntary redintegration which thus arises has two main branches; one is active, recollecting, Memory; the other is active, productive, Imagination. Both are very different from the memory and the imagination in spontaneous redintegration. They

BOOK II.
Ch. V.
——
§ 1.
Analysis of the
conceptual
process.

are governed by volition and purpose, and proceed by rejecting whatever spontaneous redintegration offers, if it does not square with the guiding purpose of the volition.

In the recollecting memory, the purpose is to reproduce exactly the order of sequence and coexistence in which the object, with which we start, actually occurred to us. Or it may be to fill up with a distinct representation a place in that order which is left empty, and the object of which is said to be forgotten. This is the phenomenon of *hunting* for a forgotten circumstance, name, or image; and it is clear from the present analysis, how it happens that we have a general and partial knowledge of what it is we want to remember, before actually remembering it. We have the empty place of it in the redintegration, and this gives us a general knowledge of what it is.

In the active and productive imagination, we are also guided by a purpose. It is to complete a picture which has never been actually presented to us, but which either might have been presented to us or to others, or which we think was the picture which must have been presented to a spectator suitably placed and suitably endowed, or in other words, the real order of phenomena and events. To discover this real order, which has never been actually presented to us, is the guiding purpose of the scientific imagination, the functions and importance of which have been so well insisted upon by Professor Tyndall.[1] Its function is to frame hypotheses and observe and bring together facts which either support or contravene them, guided by volition and basing itself

[1] On the Use and Limit of the Imagination in Science. Longmans, 1870. See pp. 18 and 51.

Book II.
Ch. V.
—
§ 1.
Analysis of the
conceptual
process.

upon facts already known. In other words its method
is Induction. An imagined order in which the facts
already known find their place and their explanation
is a scientific hypothesis; and the new facts ranged
with the old under the hypothesis are an induction,
and result in an inductive generalisation. This
whole process is by no means outside of the syllogistic method; but I reserve the explanation of the
way in which it comes under that method, till more
has been said on the nature of Propositions.[1] It is
obvious that the range of the scientific imagination
extends to the moral sciences and to history, as well
as to the physical sciences. I shall have occasion to
recur to this whole subject in the following Chapter.

The poetic imagination differs from the scientific
solely in the kind of purpose which guides its redintegrations. That purpose is of an æsthetic and
emotional character. Compared with the scientific
imagination, it is *fictive*, not restricted to the imagination of facts as they must actually have been, but
of such facts as would satisfy the desire for beauty
and imaginative pleasure.

But in both kinds of productive imagination, as
well as in recollecting memory, the order of redintegration is perceptual, the images occur as they
occur in presentative experience, as matters of fact
and events of history. The conceptual order is
abandoned for the perceptual. But it is only by
going through the moment of conception, by which
spontaneous is transformed into voluntary redintegration, that this voluntary perceptual order is
reached; and the traces of its having been passed
through are left indelibly behind it in language.

[1] Below in this Chapter, § 8.

Book II.
Ch. V.
§ 1.
Analysis of the
conceptual
process.

Language witnesses to this fact in the circumstance that all names for things and relations and events are general terms. Even proper names, the names of individuals, are originally no exception, for these have been drawn from some characteristic of the person named, such as his appearance or his home, and it is only by a tacit convention that the term is restricted to him and him alone. Proper names are an instance, and the only one in language, of terms of first intention; and they are originally terms of second intention limited by an individual application. They now *designate* without *connoting*, though originally they, too, connoted. There are, then, no such things as singular names; there are only general names limited. In *the sun*, *sun* is a general name limited by *the*. So in giving an individualised picture, as in poetic descriptions, the only way possible for doing so is by using general terms to limit each other. For instance:

> "Still as, while Saturn whirls, his stedfast shade
> Sleeps on his luminous ring."[1]

All the words here, substantive, verb, adjective, pronoun, preposition, and conjunction,—all are *general*; the only exception being the proper name, Saturn, a name given for the sole purpose of *avoiding* a name with a general connotation, but nevertheless a name applied only to that planet which has that particular place, orbit, mass, &c., which are only found *together*, that is, limiting each other, in a single instance.

Voluntary perceptual redintegration, then, depends on the conceptual order having been first adopted, and then abandoned by a further volition. The concep-

[1] Tennyson, The Palace of Art.

Book II.
Ch. V.
§ 1.
Analysis of the
conceptual
process.

tual order, too, must have been followed to a considerable extent before productive imagination can operate to much purpose. Its operations presuppose a certain work of analysis and classification, evidenced by naming, to have been previously performed. But this work is done by keeping to the first of the "two expedients;" that is, by keeping to the conceptual order. This alone is *reasoning* in its stricter sense, as opposed to voluntary imagination and memory. These supply it with its pabulum, in the shape and order of matters of fact, images represented in the same kind of order as presentative experience.

It seems as if this circumstance, I mean the priority of conception and the conceptual order to the process of voluntary imagination and memory, was the fact in Kant's mind, when he made his *reproductive* depend upon his *productive imagination*, which latter was a faculty of spontaneity (as he called it) working in and through the Categories. Within voluntary redintegration, it is clear, that conception precedes and dominates perceptual imagination and memory. But Kant, as Hume before him, seems to have passed by unperceived the immensely important distinction between spontaneous and voluntary processes (in the sense which I give to these terms). He neglected the whole domain of spontaneous redintegration, and began his account of consciousness lower down the stream, with voluntary processes; to which he prefixed a sort of porch, as it were, by his theory of the raw material of sensation or feeling offered to the understanding, in the forms of time and space. Hume's three laws of association, too, causation, contiguity, and similarity or contrast, square fairly well with the phenomena of voluntary productive imagi-

Book II.
Ch. V.
———
§ 1.
Analysis of the
conceptual
process.

nation as now described. But they offer no account
at all of the modes of association in what is really
spontaneous redintegration.

One more remark before quitting the subject. All
voluntary processes, when they have become habitual,
may be set in activity in the shape of spontaneous
processess, that is, without new effort. Spontaneous
redintegrations are not only previous to voluntary,
but come after them also. They repeat voluntary
processes spontaneously. What we have acquired, in
brain activities, we have added to our store of spon-
taneities. Thus dreams give us back images and
trains of images, which would be quite inexplicable
but for this spontaneous reproduction of conscious
states, acquired originally by means of voluntary redintegration. Dreams also show traces, and this has
always struck me as a most remarkable fact, of conceptual processes as such. This is when we seem to
know such and such a fact concerning a dream
personage, without in the least having an *image* of
that fact or of his having been concerned in it. We
seem to have a purely conceptual dream memory, as
well as one for imagery. Probably this is connected
with the memory for words.

§ 2. I now return to the consideration of reasoning in its stricter sense, the conceptual order maintained by following the first of the "two expedients," namely, redintegration with similars. And
I follow up this true conceptual process as it is performed in the direct mode of consciousness. For
in this mode, as it has been shown, there are two

PERCEPT AND CONCEPT. 309

Book II.
Ch. V.
___
§ 2.
Extension and
Intension.

ways of performing the redintegrations, one by analysing the units, the other by grouping them with their similars; only the latter of which comes forward in the primary mode. We shall thus embrace both methods, and examine them together.

In the direct mode of conception, complex objects or existents are taken as the expectant units. In analysing them they are resolved into a number of special modes of general concepts, which general concepts are reached and formed, in the first instance, by conception in primary consciousness. To analyse an object of direct consciousness is to perform redintegrations in primary consciousness. Every existent can be so analysed. A peach, for instance, is a special mode of solid extension, a special mode of colour, of hardness, of flavour, &c. The combination and interpenetration of these special modes are the existent, the peach.

But observe, this combination and interpenetration of special modes of colour, hardness, &c., is in perceptual order, not in conceptual. It constitutes the *Intension* of the peach. The peach is not the combination and interpenetration of a certain number of general concepts, but of a certain number of special modes of these, and so much specialised as to be *singular*. These singular modes are given immediately by presentative perception, by sense; and their combination and interpenetration is *shown* to us in experience, not *explained* to us in thought. We see and feel the complex object. In other words, we have the object analysed in perceptual order.

When we want to analyse it in conceptual order, we get not the singular modes given to us in sense, but the general modes distinguished and sub-distinguished, down to describing as nearly as possible the singular modes in terms drawn from their similars. Each singular mode itself is unreachable by any such description; it is merely nameable, and that by a name taken as a proper name or in its first intention. Sense gives the meaning. The *intension* is the meaning as we see it and feel it in perceptual order.

From this I distinguish the *extension*, by which I designate the meaning as we *try to grasp* it by the sub-distinctions and counter-distinctions under and between the general modes, colour, hardness, &c.; the common ground left by intersection, as it were, of all the general modes, the special modes of which combine in the peach. The extension is coextensive with the intension. It gives in conceptual order just what the intension gives in perceptual, but it does not give it equally well: the living sensation is lacking to it, and so is that minuteness of individualisation which we have no sufficient store of similars to entrap.

We have, then, before us two things, the perceptual order of intension and the conceptual order of extension, replacing the simple distinction of percept and concept. And this is the field of one of the most instructive controversies in philosophy. The scholastic doctrine known as Realism was, that the extension was the substantial reality, the intension its phenomenal image. This at least was the realism worked out in its completest shape by Duns Scotus.

§ 3. It is necessary at this point, or at any rate it will be advantageous, before continuing the analysis of reasoning in the stricter sense, to interpose some account of the great dispute between realism and nominalism in the scholastic times. This controversy, with the results reached by the thorough discussion of the questions involved in it, was the great contribution of Scholasticism to the development of philosophy, constituting the initiation into philosophy of the nations of northern Europe. Without some comprehension of it, no full understanding of the relation of the conceptual to the perceptual order is attainable. I shall base myself entirely on the recent great work of Dr. Carl Prantl,[1] as by far the most accurate, complete, and philosophical account of these matters that I know of, a work which deserves the sincere gratitude of all students of philosophy. At the same time the purposes of the present Chapter make it impossible for me to avoid giving some additional interpretation of the facts, for which Dr. Prantl is by no means to be held accountable.

Three periods must be distinguished in the history of this question, the same three which Prantl distinguishes in the history of Logic as a whole. The first extends from Isidore of Seville, in the seventh, to the end of the twelfth century; the second from the end of the twelfth century, when the newly acquired acquaintance with Byzantine and Arab authors, and with other works of Aristotle besides the Organon, began to make itself felt in Western Europe, to the end of the thirteenth or early part of the fourteenth century; and the third from the early part of the fourteenth century to the middle of the fifteenth, when Scholasticism gradually gives way to modern

---

[1] Geschichte der Logik im Abendlande. Leipzig 1855 to 1870. Four vols. have at present appeared.

science and philosophy. The chief names in the first
of these periods are those of William of Champeaux,
Roscelin, and Abélard; of the second, Albertus Magnus, St. Thomas Aquinas, and Duns Scotus; of the
third, William of Ockham, and Gerson; though, as
Prantl carefully points out, we must not consider a
few names like these as adequately representing the
rest, or exhausting the profusion of doctrines, and
systems of doctrines, which interlaced each other
with endless variety of agreement and difference.[1]

It is to the first of these periods, ending with the
close of the twelfth century, that the question between Nominalist and Realist in its strict shape
belongs. That question concerned the reality of
Universals, that is, of general terms or concepts.
Those who maintained that there were realities, outside the mind, answering to terms that were general
and not singular, were called Realists. Those who
denied it were called Nominalists. There was no
essential difference between Nominalism and what is
sometimes called Conceptualism. To exist only in
the human mind, and not out of it, was to be a *concept.* "Et tale esse in intellectu universalia habere
dixerunt illi qui vocabantur nominales."[2] There
were no doubt different ways of exhibiting the common nominalist doctrine, some of which might lay
the greatest stress on the words, and others on the
thoughts; but these differences were as nothing to
the great difference which turned on the question,—
real or not real apart from the intelligence. No

---

[1] See Work cited, Vol. II. pp. 4, 118 et seqq. 261. Vol. III. pp. 1
et seqq. 100, 330. Vol. IV. pp. 1, 147, 50, 186, 193.

[2] Quoted from Albertus Magnus : Prantl, Vol. III. pp. 94. Note
377.

PERCEPT AND CONCEPT. 313

Book II.
Ch. V.

§ 3.
The Realistic
Controversy.

nominalist ever dreamt of asserting that general terms had a meaning apart from the general thoughts or concepts of the mind. Nor is it easy to see how it could be asserted without falling at once into an empirical absolutism, in which ready-made words stood in direct relation to ready-made things. The moment words are admitted to have their meaning from convention, that moment thought is interposed as intermediary between words and things.[1]

When we come to the second period, we find this conceptual Nominalism in possession of the field, and at the same time new questions, arising from the solution of the old ones, being debated. It being admitted that singulars, individuals, were the only realities (out of the human mind or intelligence), the next question related to the determination of individuals;—what constituted an individual? This was the question of the *Principium Individuationis*. The protagonists in this discussion were Albertus Magnus and St. Thomas Aquinas on one side, Duns Scotus on the other. The former placed the principium individuationis in the *matter* of the individual thing, thereby holding, it was thought, by Aristotle; the latter in its *form;* matter and form being taken in reference to *thought.*

But the "matter" of the Thomists was not formless matter; it involved form, though not the form of thought. It was capable of being the principium individuationis only when it was already "materia *signata*," that is, matter "quæ sub certis dimensionibus consideratur." The Thomist solution consisted in saying that real presentative percepts had in them

[1] For passages bearing on the first period, see Work cited, Vol. II. pp. 8. 35-7. 118 et seqq.

the principium individuationis: it did not say in what this principium consisted. It showed where the solution was to be looked for, but stopped short of finding it. Taken *as a solution* it was empirical absolutism. Matter was individualised from the first.

On the other hand, the "form" of the Scotists was already a form of *thought*, a concept, although individual; it was a "thisness," *hæcceitas*, and there were necessarily as many *hæcceitates* as there were individual things.[1]

The Scotist solution of the question of individuality was one which started at any rate within the conditions of the former *Realist* problem; for the hæcceitas was individual, and yet at the same time it was a form of thought, of conception, an οὐσία of essence, not of existence. It crowned, so to speak, the edifice of the *essentia*, and was the last and completing *conditio sine qua non* of the positive *existentia* of the thing in question. The existentia of the individual was a *positio*, a *Setzung* as Kant would say, supervening directly upon the hæcceitas.

In this way Scotus virtually reversed the nominalism with which he began; the Thomists on the other hand remained within its results. In looking to *materia signata* for the principle of individuation they were placing reality in *percepts*, whereas Scotus placed it in *concepts*. They kept to the Aristotelian course, he reverted to the Platonic.

These were the two answers given to the question of the second period, which itself was the logical consequence of the question of the first period. That first question was—Have universals, as such, any

---

[1] Work cited, Vol. III. pp. 96-7. 100. 114-5. 119. 128. 182. 184. 189. 213. 214. 217-9.

PERCEPT AND CONCEPT. 315

Book II.
Ch. V.
—
§ 3.
The Realistic
Controversy.

real existence apart from the mind? Answer,—
No. The second question was—Are or are not universals, as objects of thought, the condition of individuality, that is, of reality, in individual things?
The same question in fact revolved, on a higher plat-
form. Only individuals, not universals, have a real
existence apart from the mind. Good. But is there
not such a thing as an individual *concept*? And is
not this the condition of an individual *thing*?

Now the *ideal limit*, as we have partly seen already
and shall see more fully presently, of the process of
conception, by intersection of concepts and rejection
of what is not common to all, is the individual. But
this ideal limit, supposing it reached, being an *individual*, could not also be an *universal*. At the limit
you cease to have an intersection of universals, and
you have, *instead of it*, an individual represented
percept. Scotus wanted to keep *both*; the *hæcceitas*
as individual concept (a contradiction in terms), on
one side the limit, and the individual percept, *hæc res*,
on the other.

The fact of the conceptual order being the obverse
aspect of the perceptual, while the nature of the connection between them was not fully perceived,
enabled the Doctor Subtilis to turn the Nominalist
position. The Thomist Nominalist doctrine, though
on the right road, was no solution of the question of
individuality. It was a mere assertion that *materia
signata* was individualised. Now Scotus did offer a
solution, though one which reversed the admitted
doctrine that presented percepts were the realities,
by making *thought* appear as the source of individuality, individuals being granted to be real.

The third period opens with William of Ockham,

and the controversies which he initiated continue
down to the end of Scholasticism. Ockham distinguished all such questions as those of *principium individuationis* as metaphysical and not logical questions;
and he took up the apparently "practical" position of
being a pure logician. He busied himself chiefly with
the *terms* employed in logic, and the proper name for
him and his followers no doubt is, as Prantl also
shows that it was among his contemporaries, not
Nominalists but Terminists. At the same time he
was compelled to go into the very same questions as
the metaphysicians, by the necessities of the logical
use of terms, the *suppositio* and *proprietates terminorum*; and accordingly, plunging into metaphysical
questions, he took up a position of antagonism both
to Scotists and Thomists, and that as well in respect
of questions relating to the existence of universals, as
of their special question of individuation.[1]

What was this position? It was nothing else
than applying his famous "razor," *Frustra fit per plura*,
&c., to the *hæcceitas* of Scotus, while rejecting the
empirical absolutism of the Thomists, by appealing to
subjective analysis of the actions of the mind. As
against Scotus he said,—Your form of thought is
superfluous. We can almost imagine he is stating
Kant's problems as he raises the question against
Scotus: "Manifestum est in intellectu esse actum intelligendi et etiam habitum, sed utrum species aliqua
prævia actui sit ponenda in anima vel non, est dubium.
Utrum etiam præter actum intelligendi sit aliquis
conceptus formatus per actum intelligendi vel etiam
sit aliquis conceptus habens tantum esse objectivum,
est dubium. Utrum etiam sit aliqua species in intel-

[1] Work cited, Vol. III. pp. 327. 334-7. 343-5. 349-60.

lectu, quæ non posset esse sine actu intelligendi, est dubium." These questions, *dubia*, he then answers in the negative by his *Frustra, &c.*[1]

As against the Thomists he said,—Not *things, materia signata,* but states of consciousness, first and second intentions, are the ultimate realities with which we have to do. " Ex quo patet, quod intentio prima et'secunda realiter distinguuntur, quia intentio prima est actus intelligendi significans res, quæ non sunt signa; intentio secunda est actus significans intentiones primas."[2] And again, " Patet ex dictis, quod tam intentiones primæ quam secundæ sunt vere entia realia et sunt vere qualitates subjective exist-entes in anima."[2] *Subjective,* that is, in the mind as their Subject and real substrate ; what we should use the very opposite word to express, viz., *objectively.* But it must be admitted that Ockham threw no new light on the special question of individuation.[3]

It was a great advance to distinguish Logic from Metaphysic, as Ockham did, and then to enquire into the action of the mind and the consequent proprietates terminorum. It was practically carrying out Aristotle's continued query, τί σημαίνει; But in the hands of Ockham's successors, this distinction became the beginning of decay, for they changed it into a divorce, and then pursued Logic apart from Metaphysic. So pursued, logic became what Prantl calls it, a rank jungle, the work no doubt of acute men, the Booles and De Morgans of logical technicality, but for all that only the better calculated to rouse Bacon's indis-

---

[1] Work cited, Vol. III. p. 335. note 758. and p. 338. note 768.

[2] For both passages see last citation.

[3] See the passage bearing on this point in Prantl, Vol. III. p. 359-60. and esp. note 815.

criminating indignation against the whole Aristotelian science, which it professed to represent.

The word *Realism*, then, in Scholasticism, has according to these remarks three different meanings; 1st, it is properly opposed to the conceptual Nominalism of the first period; 2nd, it has acquired a meaning the exact reverse of what it ought to have, in being applied to mean the Thomist doctrine as opposed to the *Formalism* of the Scotists. The Thomists held what may be called a perceptual realism, which is a contradiction in terms if realism implies reality of *universals*. The Formalists were the true realists in this sense; and seeing that the Thomists, in their search after the *principium individuationis*, stopped short at "*materia signata*," already an empirical and individual thing, the proper name which marks their position is, not realism, but empirical absolutism. 3rd. *Realism* is employed to indicate metaphysical doctrines whether of a Thomist or Scotist character, as opposed to the purely practical logic of *Terminism*, in the third period.

Nominalism, on the other hand, has but two senses; first, as opposed to the Realism of the first period; secondly, as equivalent to Terminism and opposed, as a matter of practice not of theory, to Metaphysic in the third period. In both senses it is identical with Conceptualism. It is the golden Aristotelian strain running through the whole history of Scholasticism, and working itself out with increasing clearness and distinctness by its combats with a Platonic absolutism on the one side and an empirical absolutism on the other.

Never can I concur with Dr. Prantl's refusal to recognise the independent speculative ability of the

Schoolmen. They were no mere reproducers of the new matter brought to them from time to time by Arab and Byzantine commentators. Had they been mere reproducers, there would have been no development such as we see it in the three periods just sketched. Their work was indeed left incomplete, but it was the incompleteness of living growth, of a development in two directions, modern philosophy and modern science, which again divided into various living branches. This division of the single stream of Scholasticism into the several currents of philosophy and the sciences, at the epoch of the Revival of Learning, was the euthanasy of Scholasticism and the beginning of the modern era. Scholasticism itself gave the initiatory movement to philosophy, in Ockham's distinction of Metaphysic and Logic. The movement in developing modern philosophy, though not in developing sciences, came from within, as a natural event in the living growth. The sciences had at least one root totally independent of scholasticism.

Ockham's conceptual metaphysic was the basis of the modern philosophical development, not in giving rise to a Terminist logic wholly divorced from philosophy, but in so distinguishing and connecting logic and metaphysic into a philosophical system as to leave room for physical discoveries to take their proper place within its ample lines; applying, not the *tenets* of Aristotle to support a foreign theology, but the *principles* of Aristotle to establish a reformed philosophy. The true successors of Ockham are not to be found among the Terminist logicians, but in men like Cusanus, Gerson, Telesio, Bruno, who remained faithful to philosophy as a whole. The question of

Universals, however, received its solution from Ockham; and this it was which once more placed Aristotle at the head of the philosophical development,
and determined the continuity of modern with Greek
philosophy.

We do injustice to the Schoolmen if we consider
them as the makers of the Scholastic philosophy out
of the materials brought to them from Greek sources,
without adding that they worked under the abnormal
and unfavourable condition of having a *creed*, that is
to say, an absolute philosophical system imposed upon
them to begin with, into harmony with which all
their materials, as well as all their reasonings upon
them, must be brought. Their philosophy put forth
its leaves and branches under superincumbent pressure. This condition, their tutelage under the Church,
was imposed upon them by the course of general history, the events that composed the breaking up and
resettlement of the Roman Empire, after the irrup-
tion of the Northern nations. At length the thought-
life pierced the superincumbent crust. The history
of Scholasticism is the history of that struggle of the
life of thought, partly to break through, partly to
assimilate, the absolute creed of the accepted ontological theology.

Of course the theological, as well as the philosophical, view had its partisans. But just because
both were included in Scholasticism as a whole, and
because neither can be taken as standing for the
whole, therefore we must look elsewhere for guidance
in our judgment of Scholasticism. We must look at
the result to which it led, its total outcome; and
with this alone Scholasticism must be credited. This
outcome was the division, made possible by Ockham,

PERCEPT AND CONCEPT. 321

Book II.
Ch. V.
§ 3.
The Realistic
Controversy.

of the single stream into the several branches which constitute modern philosophy, and the reconquest by philosophy of the position, in respect of the new-born sciences, originally won for it by Aristotle.

§ 4. I now turn at length to the second of the two chief ways of forming the conceptual order in direct consciousness, the way not of analysing, but of grouping a thing, as a whole, with its similars, which is the process of reasoning, strictly so called. The concept which is expressed by the name of the class of existents formed in this way is a *general* concept, if it is a provisional image expectant of its further determinations; for then it is a legitimate extension of the method of forming conception in primary consciousness, by which *Red*, *Hard*, &c., stand as shorthand expressions for special feelings of red and hard. It then belongs to the conceptual order, and all its distinctions and sub-distinctions are its *extension*, as all the corresponding existents, analysed into their components in order of perception, are its *intension*. "Man" is a general concept, the intension of which is not "all men," but all men analysed into the countless most special modes of consciousness in and as which men are objects of perception.

But if "man" is taken, not as a provisional image, but as a short-hand expression for "all men," not adding that "all men" are analysable again, but stopping short at them as ultimate existents, then it is a mere collective term, it remains an existent, a percept, and has not become a concept at all. Its analysis has still to be performed. I shall call terms of

BOOK II.
CH. V.

§ 4.
Extension,
Intension, and
Comprehen-
sion.

this sort *collectives*, and, instead of speaking of their intension, shall speak of their *comprehension*, meaning the existents which they are supposed to include.

The remarkable point about this way of taking general terms, not as concepts but as collectives, is, that intending to be thoroughgoing Nominalism it is most decided Realism,—Realism, I mean, in the Thomist sense, that is to say, empirical absolutism. It denies the realness of "man," but it asserts the realness of "this man." It leaves unanalysed the Thing-in-itself in every apparent individual. Individuals, singulars, are no doubt the only reals. But this means, not individual men or individual "things" of any sort, but individual *percepts;* anything as a percept or in order of perception, as opposed to anything in order of conception or as a concept. Percepts are the real Existents. It must be observed too, and the remark is not unimportant, that it is not strictly true to say that we perceive singulars or individuals at all. Strictly we must say, we perceive what are afterwards called singulars or individuals; afterwards—that is, after conception has supervened. The distinction between General, Particular, and Singular, is a conceptual distinction.

We may, then, define the three things, the relation of which to each other has now been given, as follows:

> *Intension* of a term;—the percepts constituent of it as a percept.
> *Extension* of a term;—the common intersection of the concepts of its definition, or of the concepts constituent of it as a concept.
> *Comprehension* of a term;—the individual percepts to which it is applicable.

PERCEPT AND CONCEPT.   323

Book II.
Ch. V.
———
§ 4.
Extension,
Intension, and
Comprehen-
sion.

For instance, the term *Triangle*. Its intension is—a certain mode of combining three lines which we know when we see it. Its extension is—a surface figure described by three lines. Its comprehension is—all the instances of such a figure, no matter what their other characteristics are.

The intension and extension of any term are, ideally, exactly coextensive; that is, it is the aim of science to make them so, whichever of the two we begin with; and in the attempt we increase our knowledge of both. For the intension of a general term, triangle for instance, is a Schema, diagram, or provisional image. Its extension is what is determined in common by two or more general terms which are its definition. As we add new general terms we determine more strictly what is common to all of them; at the same time we add a feature to the diagram which is the intension; we fill it up with a detail. The *limit* of completion of the detail of the extension is the point where the common object of the general determinations ceases to be a general, and becomes a singular, object; ceases to be a concept and becomes again a percept. And this limit may be represented in the extension of the general term by imagining the common intersection of the general terms composing it vanishing to a point, by the continued addition of new general terms, each diminishing the common intersection. The extension and intension of a term, then, vary together. And *at the limit*, the completed concept being also a completed percept, percept and concept may be regarded as obverse *aspects* of each other. Facts of this kind it was a special merit of Hegel to discover; and here it is that his analytical powers have been most fruitful.

Book II.
Ch. V.
——
§ 4.
Extension,
Intension, and
Comprehen-
sion.

He may be said to have couched our eyes for their perception.

But the comprehension of a term varies, and embraces much or little, according as that term has its extension and intension poor or rich. The fewer features there are in the extension and intension of a term, the larger number of existents there are to which that term is applicable, out of the whole number belonging to the same syngeny, or scala generum. And the greater number of features in the extension and intension, the smaller the number of existents in its comprehension, always within the limits, or compared to the individual existents covered by other terms, of the same scala. From all individual triangles which are the comprehension of the term *triangle*, the determination *rectilinear triangle*, which is an enrichment of the extension and intension, cuts off all curvilinear triangles, and lessens the comprehension.

The distinction between the comprehension and extension of general terms throws some light on the question of natural species, which has been recently so much debated. The genera, species, and varieties of organic beings are wholes of comprehension; collective wholes, of which the individuals are the parts. They are groups of individuals, which may have greater or less permanence in nature, and be more or less strongly marked out from other groups. But the notion of Type is a different matter; and this seems to me to involve some "realist" philosophical theory, inasmuch as it appears to be based on the extension of general terms, and to require the existence of the extension prior to the existence of the individuals in which it is exemplified.

PERCEPT AND CONCEPT. 325

Book II.
Ch. V.
—
§ 5.
The Logical
Categories.

§ 5. The conceptualising process, it will be observed, lands us in the Aristotelian distinction of *genus, differentia,* and *species,* which may be properly called the logical categories. These categories are not found ready made in thought, but are products of the process of conceptualising the chains of perceptual redintegration, by attention, under the law of parcimony. For this process, as it has now been analysed, carries with it and involves the arrangement of the concepts, which it produces, into classes of similars, distinguished each by some difference, answering to Aristotle's ἑτερότης τοῦ γένους, and each of these classes is again distinguishable into two or more sub-classes by another difference, that is, by a circumstance or feature belonging to one part of it, and not to another, answering to Aristotle's διαφορὰ εἰδοποιός. A class so distinguished into two, or more, sub-classes is what is called a *genus;* the circumstance which distinguishes it into two sub-classes is called a *differentia,* and the sub-class which possesses it a *species,* of the genus in question; and every *species* is said to be *defined* by its *proximate genus* and *differentia.*

This process of definition by proximate genus and differentia is evidently nothing more, and nothing less, than the process of classification reduced to its lowest terms, and in its strictest form. It draws tight, as it were, the lines by which we describe the common intersection of general terms in a term of extension. For it reduces the intersecting terms to two, and requires that there should be no general term interposable between them, or in other words, that the difference which completes the definition shall be a difference in its *proximate* genus, or term which begins the definition. At the same time it is perfectly general and universally applicable; for the difference which distinguishes *genus* from *genus* is obtained by a mere repetition of the same process as that by which *species* is distinguished from *species* within the same *genus*. The same concept which is genus to the species below it is, in its turn, species to a higher genus, until we come to one of the few determinations, different from one another, which are not themselves formed by differentiation of a higher determination except that of existence itself as a general term, the opposite aspect of which is consciousness.

The distinguishing between the two orders perceptual and conceptual, the putting categories of logic side by side with categories of perception, and resolving one into the other in all processes of reasoning,—all this, which is the foundation of Logic, is due to Aristotle. *Analytic* means, with him, the resolution of percepts into concepts with demonstration of their equivalence. To bring the theory of Syllogism and the theory of Demonstration in Necessary Matter (*Apodeixis*) to this analytical test is what justifies the title of *The Analytics* being given to the two treatises famous under that name.[1] The foundation of this analytical treatment is laid in the *Topics*, where the two orders, with the categories belonging to each, are placed side by side, and those of the one shown to be predicable of those of the other. He there first distinguishes and defines four logical categories, ὅρος. ἴδιον, γένος. συμβεβηκός. as constituting

---

[1] This is clear from many passages and particularly Anal. Post. A. 22. p. 84, a. 7. λογικῶς μὲν οὖν ἐκ τούτων . . . . ὥστε καὶ ἐπὶ τὸ κάτω.

between them the content of propositions.[1] He then
proceeds to give the perceptual categories, the well-
known ten, in which the logical categories are found,
and of which they are predicable.[2] And this relation of the two lists of categories to each other is the
foundation of the whole of Aristotle's logic, not only
of its dialectical, or general, but also of its apodeic-
tical or specially scientific branch; the theory of syllogism, or purely formal method of reasoning, being
common to both.

There is no doubt a great difference between Aristotle's list of logical categories and that list of three,
genus, differentia, species, which I now call by that
name.   Aristotle's four were, in truth, not purely
logical as opposed to perceptual categories. They
were the first step, imperfect it is true but still an
altogether decisive step, towards a purely logical
enumeration.   To him, the γένος, ἴδιον, and ὅρος, of his
list were general perceptual characters found in
nature, not characters *made* general by being fixed
on in reasoning.   But the step of distinguishing
general characteristics *in predication*, from general
characteristics in perception, was a step sufficiently
decisive to prevent logical minds from falling back
upon the old undistinguishing position, the Platonism

---

[1] Top. Λ. 4. p. 101, b. 17.   Πᾶσα δὲ πρότασις καὶ πᾶν πρόβλημα
ἢ γένος ἢ ἴδιον ἢ συμβεβηκὸς δηλοῖ · καὶ γὰρ τὴν διαφορὰν ὡς οὖσαν
γενικὴν ὁμοῦ τῷ γένει τακτέον. ἐπεὶ δὲ τοῦ ἰδίου τὸ μὲν τὸ τί ἦν εἶναι
σημαίνει, τὸ δ' οὐ σημαίνει, διῃρήσθω τὸ ἴδιον εἰς ἄμφω τὰ προειρημένα
μέρη, καὶ καλείσθω τὸ μὲν τὸ τί ἦν εἶναι σημαῖνον ὅρος, τὸ δὲ λοιπὸν
κατὰ τὴν κοινὴν περὶ αὐτῶν ἀποδοθεῖσαν ὀνομασίαν προσαγορευέσθω ἴδιον.
κ.τ.λ..

[2] Top. Λ. 9. p. 103, b. 20.   Μετὰ τοίνυν ταῦτα δεῖ διορίσασθαι τὰ
γένη τῶν κατηγοριῶν, ἐν οἷς ὑπάρχουσιν αἱ ῥηθεῖσαι τέτταρες. ἔστι δὲ
ταῦτα τὸν ἀριθμὸν δέκα, τί ἐστι, ποσόν, . . . κ.τ.λ..

which Aristotle thus left behind him. No matter that Aristotle still thought that "things" were classed *by nature* into *genera*, γένη τῶν ὄντων, and had their own specific differences from the same natural source. Not what Aristotle thought he was doing, but what he really was doing,—this was the important thing. And what he really was doing was this, he was formulating the process of volitional perception, and bringing it as a new process to bear upon and modify the process of spontaneous perception in redintegration. He could not bring in the determinations proper to general *propositions* without bringing in determinations proper to *conception*, for such propositions are only possible as expressions of a conceptual process.

Although Aristotle had risen out of the Platonic conception of χωριστὰ εἴδη, and had seen that the εἴδη were not prior to, but inherent in, the objects of experience, he yet had not risen above the conception of those objects of experience being marked off, in nature, by their inherent εἴδη, from one another; and thus his attempt was to refer each phenomenon to the ultimate class of objects in which nature had placed it, and of the laws of which nature had made it the exponent. Hence with him *genera of existents* are the true ultimates; are not cases of one highest genus, existence, or unity. The universe is with him *a collection* of genera of existents, existing in and exhibited by individual existents.

Had Aristotle seen that the highest genera were determinations and not existents, he would have given a shorter list of the various contents of propositions; we should have had two only out of the four which compose his list, namely, ἴδιον and συμβεβηκός.

For predicables, in their character as predicables, fall according to him into two classes only, those which are predicable convertibly with their subject, and those which are predicable of it but not convertibly.[1] The ἴδιον and ὅρος belong to the first class; the γένος, διαφορὰ, and συμβεβηκὸς, to the second. And this no doubt is the true ultimate distinction between predicables; either they are larger than their subject in order of extension, or they are coextensive with it. A predicate either ἐπὶ πλέον λέγεται τοῦ ὑποκειμένου, or else ἀντικατηγορεῖται, τοῦ ὑποκειμένου. It cannot be less than its subject in extension, for then it would not be predicable of the subject as a whole, but only of a part of it.

There is another sort of obscurity which invests, not Aristotle's mode only, but every mode of distinguishing and combining the perceptual and conceptual orders. It is this. Language has only one mode of expressing both orders of redintegration; for it is itself an outgrowth or an accompaniment, or rather it is at once an instrument and a product, of the conceptual order. Language throws everything into the conceptual order; you cannot speak but in general terms. "The day is fine;"—*day*, general term; *fine*, general term. Apart from proper names to some extent, as remarked above, and we may add also from interjections and imitative sounds, all language is conceptual. Hence what is really first *in genesis*, the perceptual order, is last *in analysis*, is what has to be discovered by painful interrogation of what has sprung from it, the conceptual order. First and second intentions, this apparently barbarous and

---

[1] Top. A. 8. p. 103, b. 7. ἀνάγκη γὰρ πᾶν τὸ περί τινος κατηγορούμενον ἤτοι ἀντικατηγορεῖσθαι τοῦ πράγματος ἢ μή. κ.τ.λ.

scholastic distinction, is one of the most essential in philosophy. First intentions,—the attitude of our minds in having *percepts;* second intentions,—the attitude of our minds in having *concepts.*

Nothing is more common than to hear the argument,—"Does not language itself show," &c. But this argument is hardly ever so decisive in its effect, its legitimate effect I mean, as it is supposed to be. There are many things which language may be brought in to show; many which it may not. It is evidence, but evidence which requires careful sifting. And there is one sort of things which language alone can never tell us; it cannot tell us what things it is decisive evidence for and what things not. In fact, it is evidence and not verdict.

Still, though language gives us no direct means of distinguishing these two attitudes of mind in the present case, it lends us most important aid towards it. For it gives us a class of words which may be used to express, not indeed the attitude of mind in percepts of all kinds, but at least in percepts which are most elementary, the elements in fact of "things." It gives us adjectives in distinction from substantives. Grammar in fact gives us a most important clue in this matter. For suppose that, following Aristotle, we distinguish things at large into substance, attribute or modification of attribute, and the whole formed of them,—οὐσία. συμβεβηκός, and σύνολον,—we shall find that, instead of this triple distinction, grammar knows only a double one, substantive and adjective. In fact, *substance* is superfluous. Every *synolon* may be resolved into its attributes; substance is merely an attenuated repetition of synolon, an imagined reflex of it or cause of it.

PERCEPT AND CONCEPT. 331

BOOK II.
CH. V.
§ 5.
The Logical
Categories.

Now although it is true, that all adjectives that have a meaning are general terms, and have their signification dependent on their relation to other things, yet we may employ them in preference to substantives to indicate the *feeling* which we experience in presence of a thing, in contradistinction from the thing itself. "The day is fine;"—the day makes this particular impression on me,—this would then be the sense of these words, instead of—the day is one of those days which has little or no rain. Of course I do not mean that only such words as are called adjectives in grammar can be predicates, but that predicates of categorical propositions do, as such, stand in that relation to their subjects which grammar expresses as the relation of adjective to substantive.[1] We thus get a term in its first intention in the predicate. The importance of this will be seen in treating of propositions, which is the branch of the subject next to be handled.

§ 6. The order which a spontaneous redintegration is made to take under attention for the purpose of knowledge is the conceptual order, and if expressed in words is expressed as a series of Propositions. It must not be imagined that there can be a series of concepts which are not percepts, or (in the conceptual order) a series of percepts which are not concepts; for a concept is a modified percept; and according to the strength of the volition to know, that is, to perceive the nature and the relations of the several

---

[1] This is not a new doctrine. See Mansel's Letters, Lectures, and Reviews, p. 22. The Philosophy of Language.

terms, is the distinctness of the conceptual character of the series more or less marked. If this volition is weak, we still have the conceptual character present, though not marked; it is preserved and evidenced by the language itself which is a conceptual product. Of this character are the propositions of common conversation. If again the volition is to have the perceptual character of the terms brought out, to have a series of images in the mind for their own sake, then we have a series of propositions of a perceptual character distinctly marked, such as are found in works of poetic and scientific imagination, and of narrative. We are concerned here only with the first of these three classes of series of propositions; with that in which the conceptual character is strongly marked; with that kind of propositions which is employed in reasoning and in logic.

Every categorical proposition of this kind begins with a percept changed into a concept, and ends with a concept which can be changed into a percept again. The first of these is the *subject*, the last the *predicate*, of the proposition. Between them comes the *copula*, which marks the point of transition from one to the other, taking the place of the point of change between the *differents* of which the spontaneous redintegration consisted. The copula has no content of its own, but it bears a double character, in virtue of the volition to know, which puts it in the place of the point of change between differents; either it is affirmative or else it is negative, signifying either that the predicate does coalesce into a single image with the subject, or else that it does not. If it does coalesce, we have an affirmative proposition, an enriched image, the subject combined with the predicate. If it does not coalesce,

PERCEPT AND CONCEPT. 333

Book II.
Ch. V.

§ 6.
Propositions
and Reasoning.

we have a negative proposition, by which our knowledge is enlarged but not our image enriched; a way of error has been explored and marked.

This proceeding, which is called *Judgment*, judgments being what are expressed by propositions, is the whole of the reasoning process. There is nothing more than this process of judgment in the whole process of reasoning, in various forms it is true, and variously combined; but still no more than this process of judgment.

Hypothetical judgments, one of the two main kinds of judgments, the other being the Categorical, are judgments expressed by propositions where a proposition (or propositions) takes the place of the subject, and a proposition (or propositions) takes the place of the predicate; and the place of the copula is filled by the formula "if—then," "if the subject, then the predicate."

An affirmative categorical proposition has a result, an enriched image; this may be expressed either by a new *term*, or by the proposition which affirmed it. Hypothetical propositions are merely forms either of expressing complex terms in the shape of the propositions which affirm them, and for combining these terms in a new proposition, or, in the case of negative categorical propositions, of employing their negative result as the condition of a new proposition.

Negative hypothetical propositions there are none. There are hypothetical propositions in which the consequent and the antecedent, either or both, may be negative, but the bond between the two is always affirmative. The reason of this is, that, time being adopted as the mould into which the judgment is cast, a separation of the two complex terms, which are its antecedent and consequent, from each other in time being assumed, then, unless an inner connection between them was asserted by the proposition, there would be no proposition at all. The notion of causal connection, of dependence, or condition, therefore, lies at the root of judgments which have the hypothetical form, and such judgments are always affirmative. It is a consequence of this separation of terms in time, that the syllogisms which are built upon hypothetical propositions are syllogisms which are without middle terms. There is no coalescence of terms, and no use for a means of coalescence or middle term.

To return, then, to the plain categorical proposition, which is the foundation of the whole, and to which all other varieties may be reduced in the last resort. The percept fixed upon to serve as the subject in an affirmative categorical proposition is transformed into a concept by that act. It becomes an *expectant*, yet without ceasing to be an *existent*. It is a determinate percept expecting a further determination, which the predicate will give it. When this predicate is offered by spontaneous redintegration it is in the shape either of some attribute of the subject, or of some relation in which it stands to other percepts. In either case it is an addition which is not merely attached to, but one which modifies, the subject, an addition which works a change in the subject, and makes it different from what it was before to the reasoner. The predicate stands to its subject in the same relation as adjectives stand to substantives. The coalescence of images in redintegration, which is the substratum of reasoning, is the fact which determines this adjectival, attributive, or modifying, relationship of predicate to subject. Unless we suppose

PERCEPT AND CONCEPT. 335

Book II.
Ch. V.
––––
§ 6.
Propositions
and Reasoning.

consciousness to have come into existence equipped with a furniture of classes and sub-classes ready for application in processes of reasoning, we must explain the perceived nexus between phenomena from the nexus in spontaneous redintegration; the characteristic, however, of the nexus between phenomena, even in cases of greatest difference, namely in conception, where the natural or spontaneous order is arbitrarily broken for the sake of knowing,—the characteristic of this nexus is, that it is not a nexus between *separates*, but a coalescence between *differents*, where the first member is carried over into the second and reappears in it modified. We do not lose sight of the *rose*, when we predicate that it is red; but we have before us *a red rose*.

Accordingly, it is not a final account of the import of predication to say, that "the rose is red" means "the rose belongs to the class of red things." For we have to explain how we come by our class of red things, how we come by our notion of class at all. Classes are formed by means of predication; predication does not take place by means of classes. Classification is a result and not a condition of predication.

But what is the origin of the mistake, the fatal mistake, of considering classes as a condition of predication? It is the stopping short in reflective analysis of consciousness, and, instead of analysing its direct into its primary mode, beginning with "things" as objects of direct consciousness, as if they were the ultimate data of experience; a distinction fully drawn out in the present Chapter. And a class of red things is as much an object of direct consciousness as a single red thing is, although its members

are scattered up and down the world, and not concentrated in one spot as the qualities are which compose a single thing.

This starting with "things" as the ultimate is the
basis of all separatist logic.   For instance, we find it
so in the extremely able and important work of Professor Jevons, where it is the basis of his main and
distinctive logical doctrine, "the substitution of
similars."   He there says:[1] "The simplest and most
palpable meaning which can belong to a term consists
of some single material object, such as Westminster
Abbey, the Sun, Sirius, Stonehenge, &c.   It is probable that in the earliest stages of intellect only concrete and palpable things are the objects of thought."
"Things" being supposed to be given originally in
this separate condition, it follows that the first principle of reasoning must contain an explanation of
their combination in thought.   Accordingly, Professor Jevons furnishes us with such a principle, in the
Substitution of Similars.   I select what seems to be
the chief statement of this principle.   "The fundamental action of our reasoning faculties consists in
inferring or carrying to a new instance of a phenomenon whatever we have previously known of its
like, analogue, equivalent or equal.   Sameness or
identity presents itself in all degrees, and is known
under various names; but the great rule of inference
embraces all degrees, and affirms that *so far as there
exists sameness, identity or likeness, what is true of one
thing will be true of the other*."[2]   Of course I am not
going to argue against this principle.   I mention it

---

[1] The Principles of Science.   London, 1874.   Vol. I. p. 29, and
see again p. 31.

[2] Same place, Vol. I. p. 11.

only in order to show its filiation from the separatist theory of perception, the assumption of palpable "things" as the perceptual ultimates. And this I have mentioned only because I hold it to have originated the mistake of considering classes as a condition prior to predication, to which subject I now return.

The effect of this mistake has been to substitute the Comprehension of terms for their Extension, and, as a consequence of this, to burden the world with an useless two thirds at least of Formal Logic. For when general concepts, instead of being taken as the units of thought, were taken as *collectives*, that is, as analysable into the individuals which made up their *comprehension*, (instead of analysable into the qualities which made up their intension, which would have brought them back at once to the test of immediate experience), then of necessity arose at once a system of quantified predication. For the *quantity* of propositions, universal, particular, and singular, according as the subject contained *all*, or *some*, or only *one*, of the collective term, became the most prominent feature in thought. The whole doctrine of the Figures and Moods of syllogism, and of Opposition and Conversion of propositions, followed necessarily on the distinction in their quantity. It was but a logical and legitimate extension of this same erroneous principle, when Sir William Hamilton introduced a quantified predicate as well as a quantified subject. And the denaturalisation of the meaning of the copula, if not a consequence of quantified predication, was yet a closely allied error, and one which harmonised perfectly with the other.

§ 7. It will be worth while to interpose a few remarks on the scope and origin of Formal Logic in the hands of Aristotle. It had two originating circumstances; first, the prevalence of oral discussion and argument on all kinds of subjects, second, the importance attributed to Dialectic by Plato as a means of discovering truth. Aristotle, true metaphysician as he was, and bent on doing thoroughly the work that lay immediately before him, set himself to analyse this process at once of thought and of argument with others. Let us see, he must have said to himself, what it is that we are doing when we reason with ourselves or with others. The result was to distinguish, as we have seen above, the logical categories, ὅρος, ἴδιον, γένος, and συμβεβηκός, from the perceptual categories, his well known ten, the γένη τῶν κατηγοριῶν, ἐν οἷς ὑπάρχουσιν αἱ ῥηθεῖσαι τέτταρες.[1]

The functions of Plato's εἴδη were now distributed. With Plato they were causal as well as analytical of the phenomena, τὰ πράγματα. They were γένη τῶν ὄντων, as well as logical, as well as perceptual, categories; three functions were performed by them. With Aristotle the causal function was left in the πράγματα themselves, but as analysed into the γένη τῶν κατηγοριῶν, while the analytical function was attributed to the logical categories. Instead of there being εἴδη παρὰ τὰ πράγματα, as with Plato, there were now εἴδη κατὰ τῶν πραγμάτων, predicable of the phenomena by reason of features found in the phenomena; features of the extension, or logical order, because expressing features of the intension, or perceptual order. Two functions out of Plato's three were thus accounted for; but the third remained, the function

[1] Topica, A. cap. 1 to 9 inclusive, p. 101 b. 11, to p. 104.

of the εἴδη as γένη τῶν ὄντων, the function of explaining why the πράγματα themselves were what they were; and this was a problem which Aristotle left unsolved, and which, it may be added, is still the problem of modern science and philosophy.

But this great result of genuine metaphysical analysis was equally applicable to phenomena considered as facts and laws of nature, and to phenomena considered as object-matter of discussion and argument. In the first case, it had a restricted application, namely, to facts which were καθ᾽ αὑτὰ, καθόλου, and κατὰ παντός, to what Aristotle called *necessary* matter,[1] by which he meant things and events in which was involved a *real* and *regular* causation; and this causation it was the purpose of Apodeictic to trace. In other words, Dialectic as applied to the demonstration of laws of nature became and was *Apodeictic;* and the Posterior Analytics of Aristotle is accordingly a treatise on the conditions and method of discovering and demonstrating the laws of nature.

The major premisses of apodeictic syllogisms were propositions stating definitions, that is, the οὐσία, the genus and differentia, of the term which was the subject of the proposition; this definition, the διαίρεσις of the term defined, was also its analysis, ἀνατομή, in perceptual order of intension; and therefore it contained the *cause* why the particulars, embraced by the term defined, had the properties expressed by the definition.[2]

---

[1] Anal. Post. A. cap. 6. p. 71 b. 5.

[2] Id. B. cap. 14. p. 98 a. 1. Πρὸς δὲ τὸ ἔχειν τὰ προβλήματα ἐκλέγειν δεῖ τάς τε ἀνατομὰς καὶ τὰς διαιρέσεις, οὕτω δὲ ἐκλέγειν, ὑποθέμενον τὸ γένος τὸ κοινὸν ἀπάντων, οἷον εἰ ζῷα εἴη τὰ τεθεωρημένα,

This definition was the middle term of the syllogism, and its definitum the major term; and in the major premiss the major term was predicated of its definition, the two terms being convertible. And the demonstration consisted in bringing the minor term *under* the definition of the major term, and therefore under the major term itself.[1] Thus was obtained an apodeictic syllogism in Barbara, *e.g.* :

> Trees that dry their sap are trees that shed their leaves,
>
> Broad leaved trees are trees that dry their sap,
>
> ∴ Broad leaved trees are trees that shed their leaves.

The definition in the middle term contains the *cause* of the shedding of the leaves. It is not only its *causa cognoscendi* but also its *causa existendi;* and in fact the *coincidence* of these two *causæ* is one essential characteristic of Apodeictic.[2]

Apodeictic syllogisms, then, expressed *at once* a fact and the reason for it; or, in other words, demonstrated a fact *by* pointing out its cause ; this cause being found in its perceptual analysis. Facts alone, without their reasons, were not the object-matter of apodeictic, however complete or exhaustive the enumeration of instances might be. "All swans are white," even supposing it to be strictly true, could not be a major premiss in apodeictic. The fact that the planets move round the sun in ellipses

---

πᾶσα ταυτὶ ζῷῳ ὑπάρχει. * * * δῆλον γὰρ ὅτι ἔξομεν ἤδη λέγειν τὸ διὰ τί ὑπάρχει τὰ ἑπόμενα τοῖς ὑπὸ τὸ κοινόν, οἷον διὰ τί ἀνθρώπῳ ἢ ἵππῳ ὑπάρχει. κ.τ.λ.

[1] Anal. Post. B. 17. p. 99 a. 1-4.   Also p. 99 a. 16-29.

[2] See Anal. Post. A. 2. p. 71 b. 19.

PERCEPT AND CONCEPT. 341

Book II.
Ch. V.
§ 7.
The Logic of
Aristotle.

would not be a major premiss in apodeictic; but,
introduce gravitation into the definition of planets,
and you would get such a premiss at once.

In apodeictic, then, we stand on a very different
footing from the formal syllogisms of quantified predication set forth in the Prior Analytics. We are
moving in concrete object-matter, just as in Dialectic,
but in a specialised, restricted, kind of object-matter.
The analysis of terms in order of extension and in
order of intension is what we have to do with; not
with their statement in propositions in order of comprehension. The syllogistic method belongs to both
dialectic and apodeictic, but formal syllogisms in
order of comprehension suit dialectic only. Even
formal syllogisms in the First Figure, which Aristotle
says is the most akin to science, μάλιστα ἐπιστημονικὸν,[1]
are inadequate to apodeictic logic, for their major
premisses give no indication, by their form, whether
they express a mere universal fact, such as "all
swans are white," or an ἄμεσος ἀρχὴ of demonstration,
a true καθόλου, a definition founded on perceptual
analysis. The obtaining of the ἄμεσοι ἀρχαὶ is the
real difficulty and problem in Aristotle's Apodeictic,
as it is in modern science. These are obtained, as
Aristotle shows, by Induction, ἐπαγωγή.[2] And I shall
presently make it evident, that this process, too, falls
under the general syllogistic method.

We can now see clearly how it was that *collectives*
were brought into formal logic. It was because logic
was formulated to bring to rule, and therefore to be
immediately applicable to, any and all sorts of oral
discussion. Now oral discussion could not move at

[1] Anal. Post. A. 14. p. 79 a. 17.
[2] Id. B. 19. p. 99 b. 15.

all without using the phrases "all" and "some," or their equivalents, because it dealt with the ready-made objects of direct consciousness. Hence, if Aristotle wanted to formulate oral discussion, he necessarily had to formulate the modes of speech which oral discussion employed. This is the origin of quantified predication in formal logic. It mattered not that the analysis on which the whole was founded, that of the logical and perceptual categories, showed or would have showed, if pushed to its legitimate issue, that collectives as general terms were a self-contradiction; they had to be formulated in logic, because they were forms of common speech.

We may see, then, that Aristotle's syllogistic system was an instance of what I remarked in Chapter I.;[1] it was a system which acted as a means of testing and verifying an analytical theory. The system held water, was practically sound; therefore the principles upon which it was based, and from which it necessarily followed, under certain proper restrictions, were true; for had they not been so, the system would not have held water.

But unfortunately the system being admitted to be true, and true as a test of the correctness of all reasonings and discussions, was taken as a result valuable in itself, valuable as a means of carrying on discussions, and even of discovering truth in the phenomena of nature. It was considered as something to be worked at, elaborated, carried to minute complexity. It was forgotten that one half of its work was done once and for all, in that it had served as a test of the analytical principles which gave it birth, and that, in the other half of its work, namely,

[1] At page 32.

PERCEPT AND CONCEPT. 343

Book II.
Ch. V.

§ 7.
The Logic of
Aristotle.

regulating oral discussion, its merit consisted not in
its complication but in the very reverse, in its simplicity and ready applicability. No one ever did or
ever will use formal logic as a primary method of
reasoning or argument; but only as a test of primary
methods. If it were a primary method, then no
doubt the more there should be of it, the better it
would be; but being only regulative and testing, the
less the better, provided the work of regulation can
be efficiently performed.

§ 8. Here is the place to solve two cognate difficulties which have given rise to objections against
syllogistic logic, first, the question of the relation of
induction to syllogism, secondly, the supposed *petitio
principii* involved in syllogisms. Both are solved by
the distinction now explained between syllogisms
which are composed of quantified propositions, the
terms of which are collectives, and syllogisms which
treat concepts as units of proposition, the terms of
which are analysable by their *intension*, not by their
comprehension. The objections raised against syllogism on both difficulties seem to me unanswerable,
so long as we understand, by syllogism, syllogisms
composed of quantified propositions; but are easily
disposed of when we take syllogism in its true sense,
as syllogistic process treating concepts as units.

First, then, as to the relation of induction to
syllogism. Aristotle maintains in the concluding
Chapter of his Posterior Analytics,[1] that the ultimates, the ἄμεσοι ἀρχαὶ or major premisses, of apo-

[1] Anal. Post. B. 19. p. 99 b. 15 et seqq.

Book II.
Ch. V.
───
§ 8.
Induction and
Syllogism.

deictic syllogisms are gathered from experience by an exercise of intelligence, νοῦς, working by the method of induction, ἐπαγωγή. Induction, then, is a process outside of, and prior to, apodeictic syllogism. But in the Prior Analytics[1] he brings induction *under* syllogism, by assigning a syllogistic form for the inductions by which these apodeictic ἀρχαί are obtained. Which, then, of the two is really prior, and the condition of the other? Does induction move by the laws of syllogism, or does it supply the major premiss in *every* syllogism used in science?

Grote, the distinguished historian of Greece, in his extremely valuable work on Aristotle, argues strongly against the validity of Aristotle's Inductive Syllogism, concluding his argument thus: " We thus see that this very peculiar Syllogism from Induction is (as Aristotle himself remarks) the opposite or antithesis of a genuine Syllogism. It has no proper middle term; the conclusion in which it results is the first or major proposition, the characteristic feature of which it is to be *immediate*, or not to be demonstrated through a middle term."[2] According to this, then, induction must be held as an independent process and a prior process to syllogism; syllogism a posterior process dependent on induction. And an induction is representable as a syllogism only by the assumption that the individual cases examined are all, or in some way equivalent to all, the cases contained under the class name which is taken as the middle term.

In the first place, it must strike every one as curious, that Aristotle himself should have made the

---

[1] Anal. Prior. B. 23, p. 68 b. 15 et seqq.
[2] Aristotle, Vol. I. p. 268-274, and the note.

PERCEPT AND CONCEPT. 345

Book II.
Ch. V.
—
§ 8.
Induction and
Syllogism.

same remark as his critic makes in refuting him; the proof of the invalidity of the inductive syllogism consisting simply in exhibiting and enforcing what the introducer of the syllogism himself remarks concerning it. It is obvious on the face of the passage in Aristotle, that there is a sense in which syllogism is the *opposite* of induction, and another sense in which they *agree*. I cannot help wishing that Grote had devoted his great powers, not to refutation, but to discriminating these two senses, and thus clearing up a point which Aristotle had left obscure.

Again, when Grote remarks, " Accordingly, Aristotle directs us to supplement these premisses by the extraneous assumption or postulate, that C the minor comprises *all* the individual animals that are bileless,"[1]—although he quotes the words δεῖ δὲ νοεῖν τὸ Γ, κ.τ.λ.,—he does not observe that νοεῖν is the technical word for the mental function employed in induction, ἐπαγωγή,[2] and that therefore the premisses cannot be said to be supplemented by an *extraneous assumption or postulate*, because this supplementing is precisely the *differentia* of the inductive from the ordinary syllogism. If there is an assumption at all, it belongs to the induction ; and if induction does not warrant the extension of the term C to all the individuals, it is not warranted at all. What Grote calls an " extraneous assumption" belongs to the inductive syllogism, not *qua* syllogism, but *qua* induction ; belongs to its *inductive* character.

My solution is the following. So long as we have in view quantified syllogisms only, syllogism is nothing more than a setting forth of the results of

---

[1] Aristotle, Vol. I. p. 272.

[2] Anal. Post. passage above cited.

PERCEPT AND CONCEPT.

induction. But what if we take syllogism as a process of reasoning by unquantified propositions, each term of which is taken as an unit and is analysable by its *intension*? Let us turn to the account above given of induction as the method of scientific imagination.[1] Two things must be distinguished; there is induction a *method* of investigation, and there are the several acts of observation, the facts observed brought together one by one, from which the method takes its name *induction*, ἐπαγωγή, or fact upon fact. The observation of a single fact is not, but the observation of two facts together is, an induction. Induction as *method* depends on the use of hypothesis by imagination; but the analysis of that method is into *acts* of induction, *i.e.*, acts combining two severally observed facts. It may be usefully remarked also, that Aristotle's νοῦς, the mental function by which ἐπαγωγή is performed, is the equivalent, in his theory, of *hypothesis* in the modern theory of the method of induction. The modern theory and Aristotle's are therefore in complete accordance with each other.

Now I say that the combining two severally observed facts is a process of reasoning which is syllogistic in its nature; it cannot be expressed but by a complex proposition, which, being complex, may be analysed into the three propositions of a syllogism. Suppose I have a piece of Iceland spar before me. I observe that it is a crystal, observation 1; I then observe its double refraction, observation 2; I then say, this is a crystal having double refraction. The Iceland spar is the middle term of a syllogism in Aristotle's Third Figure, but one which consists of unquantified propositions.

[1] Above in this Chapter, p. 305.

This piece of Iceland spar has double
  refraction   .   .   .   .   B is A
This piece of Iceland spar is a crystal . B is C
Therefore this crystal has double refraction   .   . ∴ C is A.

Of course all that is made out by this syllogism
is this fact and no more, that an individual piece of
crystal has double refraction. But observe, it not
only records the result of a double observation, but
it expresses in terms and propositions the making
and the combining of the observations themselves.
The distinct expression of the observations follows
from the observations themselves being distinct while
being made. The syllogism does not register (to use
J. S. Mill's word) the induction only, as a whole,
that is, its *result*, but it registers each observation in
it as it is made. It is immediate to the *observations.*
One distinct observation cannot be combined with
another distinct observation except by a mental process moving by the Postulates of Logic. I conclude,
therefore, that induction is a case of syllogism, or
general syllogistic reasoning, when you take both syllogism and induction in their true sense, and analyse the lowest and simplest acts of each.

The same distinctions and the same criticism are
applicable to deduction as to induction. There are
deductive *acts* and a deductive *method.* Every syllogism is a deductive *act;* the name simply expresses
the relation which the conclusion bears to the premisses. But deduction as a *method* rests on a previous
process or method of induction, and begins from its
result, as a major premiss. It then follows a course,
the direction of which is the opposite to that taken

by induction, coming down to particulars from generals; not however to *the same* particulars as those upon which it was founded, but to new ones which have the same features as the original ones, but differently combined, or in different circumstances. Deduction also is syllogistic; not identical with it, but a case of it. The identification of it with syllogism is tantamount to including induction under it; and then, since all major premisses, in Aristotle's Apodeictic at least, are obtained by induction, deduction is made to obtain (illogically) its own major premisses. Neither deduction nor induction is the whole of syllogism; but both are cases of it. Inference in every shape is syllogistic, and there is inference in both the methods because there is inference in the acts composing both.

And now comes the question as to *petitio principii* in all syllogisms. Here again, supposing syllogisms of quantified propositions to be spoken of, I cannot but agree with what J. S. Mill says, in his Chapter on the Functions and Value of the Syllogism, "It must be granted, that in every syllogism, considered as an argument to prove the conclusion, there is a *petitio principii.*"[1] To take Mill's instance,

> All men are mortal,
> Socrates is a man,
> ∴ Socrates is mortal;

plainly, until we know that Socrates is mortal, we do not know that All men are mortal, and to assert this as the ground of proof that Socrates is mortal is to assume what you profess to prove.

But the same answer must be made here as in

[1] System of Logic, Vol. I. p. 205. 6th edition.

the former case. The objection applies only to quantified syllogisms. If we take the terms as units analysable into their *intension*, the *petitio principii* disappears. Replace "all men" as a collection of individuals by "man" meaning, say, "existents that are rational and corruptible;" understand by "mortal" the feature of "sooner or later ceasing to live;" and the syllogism stands as the expression of a double observation,

To be rational and corruptible is sooner or later to cease to live,
Socrates is rational and corruptible,
Therefore Socrates sooner or later will cease to live.

It is not requisite that the case of Socrates should have been examined in order to our assurance that *composita solvuntur*. The appearance of its being requisite comes solely from the form, by *collectives* being used as terms; for "Socrates" is requisite as a component of the term "all men."

Quantified syllogisms are not the legitimate expression of the syllogistic process. The separatist logicians have here a clear victory over Aristotle. And since he also represents the syllogistic process as an account of the mental process in all reasoning, which it truly is, the separatist logicians seem to have discovered a flaw inherent in the process of reasoning itself. Hence their inclination to set up an Induction which is not syllogistic, as the rival of syllogism, which they characterise as *deduction*, and to establish that all reasoning, all real inference, is induction.

Mill classes together Syllogism, Ratiocination, and Deduction, opposes them to Induction, and then ascribes the whole of real reasoning, the whole illative force or cogency of it, to induction; syllogism, ratiocination, and deduction, having the functions of registering and interpreting its results. Inference resides not in deduction, but wholly in induction.[1] Induction is not so named because it is a process which includes the *observation of new facts;* (so at least I understand Mill); but because it is a process which *infers new truth* from facts taken as given. "In every induction," he says, "we proceed from truths which we knew, to truths which we did not know; from facts certified by observation, to facts which we have not observed; and even to facts not capable of being now observed; future facts, for example; but which we do not hesitate to believe on the sole evidence of the induction itself."[2]

Induction, then, according to Mill is inference, and yet is not syllogistic. But how inference can be other than syllogistic, if it is a sequence of thought, or conception, or reasoning, expressible in propositions, I am unable to see. There is an inferential and syllogistic process underlying the *methods* both of induction and deduction, because it enters into every act of combination composing both; and these two methods may be properly contrasted as the two parts, former and latter, of the whole process of discovery and proof, induction leading up to general truths, and deduction down to new cases under them. And the whole being thus an inferential or syllogistic process in both its parts, the inductive method is farther distinguished from the deductive, first by its

---

[1] System of Logic, Vol. I. pp. 208, 216, 218, 220, 229, 236, 247 (most important passage). 6th edit.

[2] The same, Vol. I. p. 185. See also pp. 321, 343.

PERCEPT AND CONCEPT. 351

Book II.
Ch V.

§ 8.
Induction and
Syllogism.

being guided by hypothesis, and then by its including the observation of new facts, facts not comprised in framing the hypothesis, along with the old facts which were so comprised, and inferring finally a new general proposition expressing the combination of the new with the old facts.

After reading Mill's theory subordinating deduction to induction as its mere interpretation, and attributing the whole inferential force to induction, it will be instructive to turn to another distinguished separatist logician, and find Professor Jevons taking the opposite line, subordinating induction, and treating it as "simply the inverse employment of deduction."[1] This, he effects, too, on the same condition, so to speak, on which Mill effected the very opposite, the condition, namely, of first emptying induction of its characteristic, which is its going on to the *observation* of new facts. By induction he means a process of inference, not including new observations. "Neither in inductive nor in deductive reasoning can we add a tittle to our implicit knowledge, which is like that contained in an unread book or a sealed letter."[2] Again, "The process by which they" [the general truths to be ascertained by induction] "are reached is *analytical*, and consists in separating the complex combinations in which natural phenomena are presented to us, and determining the relations of separate qualities. Given events obeying certain unknown laws, we have to discover the laws obeyed."[3] He supposes Nature to give us phenomena in batches, as it were, constituting problems ready stated for us.

[1] Principles of Science, Vol. I. Preface, p. vi. also pp. 13-15.
[2] The same, Vol. I. p. 136.
[3] The same, Vol. I. p. 140.

" The effects are now given and the law is required."[1]
Several other passages contain the same doctrine.[2]
In the last of these passages, indeed, Professor Jevons
speaks of observation of facts as one of the three
" steps in the process of induction." But it is only
observation of " the particular facts under consideration," the " facts in question," and which we begin by
" being in possession" of. Professor Jevons, then,
while inseparably connecting induction with deduction, does so only by divorcing induction from *new
observation*. But by adopting my distinction it is
made clear, that the *method* called induction includes
both inference and new observation; while each
separate *act* of induction is expressible only as a proposition, and therefore, being also complex, contains
an inference, namely, the combination of two facts
not originally seen together.

In fine, the syllogistic process is not identical
either with deduction or with induction, but is common to both. It is a process the principles of which
are involved in every judgment and every proposition; and therefore it is that it is inseparable from
methods of reasoning, by whatever names they may
be called, or whatever their other differences. As a
process it is the continuation of the process of judgment, one judgment in combination with another,
just as terms are combined in judgments. What the
copula is in categorical propositions, that the middle
term is in categorical syllogisms. Syllogism is therefore a far more abstract and elementary process than
either induction or deduction, though it is common
to both, and serves to distinguish both alike from

[1] Principles of Science, Vol. I. p. 140.
[2] The same, Vol. I. pp. 167 168. 250-1. 262. 307-8.

PERCEPT AND CONCEPT. 353

BOOK II.
CH. V.

§ 8.
Induction and
Syllogism.

simple observation or perception. The mark which distinguishes reasoning from perceiving is specially characteristic neither of deductive nor of inductive reasoning, but consists in the act of attention which modifies percepts into concepts, and in its expression, the Postulates of Logic.

§ 9. But to return to the main thread of our analysis. The copula means coalescence or non-coalescence. This is the true and also the Aristotelian doctrine. It is untrue and anti-Aristotelian to hold that the (affirmative) copula means either identity, equality, or convertibility. The word " is" in propositions cannot have the same meaning as the sign =, in mathematic. Mathematic has one predicate only, with its contrary, in its whole range of demonstrations; it predicates equality or non-equality, it affirms or denies equality, between all the quantities, whether of number or magnitude, which come before it. For its definitions, indeed, for its classifications of *kinds* of quantity, it seeks the aid of logic, and the mathematician becomes a logician; but in its demonstrations, in its calculations, it has no predicate but equality. This is to say, in other terms, that to measure and to calculate is the purpose of mathematic. Accordingly, having but one predicate, there is no harm but rather good, (since it facilitates calculation by keeping the two terms, compared in respect of equality or inequality, each for itself before the mind), in throwing this one predicate into the copula, and using =, where the general reasoner or logician uses " is." But logicians, whose purpose is not to

measure or calculate, but to analyse all the relations
and modes of things, in respect of their kind and the
manner of their relationship, should at least, before
adopting this device and importing the predicate of
equality into their copula, have paid more attention
than they seem to have done to Aristotle's distinct
denial of the possibility of so doing and to the whole
structure of his logical system.[1]

Propositions which assert identity or equality
between subject and predicate, (in which therefore
the terms are convertible), e.g., All A is all B. are
really not one proposition but two; the two being
All A is B, and All B is A;—which is an old remark.
It is indeed true, that there are propositions whose
terms are convertible. But there is no mark in the
form of the proposition itself to indicate whether the
terms are so or not. And this being so, every proposition must, in Logic, be taken to mean the least that
it can mean; and the distinction between concrete
propositions whose terms are, and those whose terms
are not, convertible is not a distinction in pure Logic,
but is the first distinction of *Method* based upon Logic.
Those who adopt identity, equality, or convertibility,
as the meaning of the copula, are adopting the distinctive mark of Aristotle's method of Apodeictic, in
which the major premisses are definitions, and ex-
tending it to embrace Logic with all its methods, as
a whole. They thus leave Logic itself in the air, baseless, that is, with its principle and method unanalysed,
and yet capable of analysis and demanding it. Nor
yet in doing this do they escape from the ambiguity
just mentioned, which they signalise in the use of

[1] De Interp. 7. p. 17 b. 12. And see Anal. Prior. A. cap. 2.
p. 25 a, 7.

PERCEPT AND CONCEPT. 355

Book II.
Ch. V.
———
§ 9.
Meaning of the
Copula.

the word "is." The same ambiguity attaches also to the sign =. Thus Professor Jevons says: "I conceive that the sign = always denotes some form or degree of sameness or equivalency, and the particular form is usually indicated by the nature of the terms joined by it."[1] If you adopt a copula meaning identity or equality, you will of necessity require some distinctive form to show when only a partial identity is intended. And accordingly Professor Jevons in such cases repeats the subject in the predicate, so as to divide the whole natural range of its meaning and restrict it to a part equal to the subject. For instance, where ordinary logicians say "Man is mortal," or A is B, Professor Jevons says "Man is man mortal," or $A = A B$.[2]

Aristotle, although he made use of collectives and quantified his propositions, yet had not so far lost sight of the true analysis of conscious thought as to mistake the meaning of predication, and quantify his copula by making it mean equality; he did not take either this step, or even that of quantifying his predicates, towards the final consummation of mathematicising logic. But the introduction of quantity at all into propositions, (owing as I have shown to collectives being substituted for concepts, and comprehension for extension, and this to defective analysis stopping short at objects of direct consciousness), has worked like a leaven in the science which Aristotle founded, and we see the fruits of it at the present day in the writings of able mathematicians, like the late Professors Boole and De Morgan, and even in the work above cited of Pro-

[1] Principles of Science, Vol. I. p. 19.
[2] The same, Vol. 1 p. 47. 48.

fessor Jevons, notwithstanding his distinct belief that
he is reversing Boole's procedure, and subsuming
mathematic under logic, instead of logic under mathematic.[1]

No system of logical formulas, however completely,
minutely, and accurately adapted to represent the
classes and distinctions of things and of relations
discovered in objective phenomena, can supply—I do
not say a means of discovering—but a test of the
truth of discoveries when made, *unless* the logical
formulas are also, and independently, expressions of
the subjective process of thought employed in the
discovery, *unless*, therefore, they give back *primarily*,
not the distinctions found in the things, but the steps
and nexus of steps in the progress of the mind towards finding them. Except from this independence
of origin, logical laws have no *testing* power at all.
They are mere repetitions of, stamps impressed by,
the facts which they are supposed to test. They are,
in other words, special sciences repeated, mathematic
in guise of logic. The special sciences, and mathematic at their head, have an objective logic of their
own, drawn from the distinctions of the object-matter
which they investigate. We speak, for instance, of
the Logic of Political Economy. It is a form of expression suitable to the truths of a science while yet
in process of being systematised.

Now I do not deny that an artificial system of
logical formulas may be invented, founded it may
be on the principle of identity or equivalence of Aspects, namely, of definition and its definitum, as in
Aristotle's Apodeictic, which is a reflective principle,
very different from the Postulate of Identity, A is A;

[1] Principles of Science, Vol. I. pp. 83. 130. 173.

PERCEPT AND CONCEPT. 357

Book II.
Ch. V.

§ 9.
Meaning of the
Copula.

nor that such a system may be capable of rendering,
and rendering better, all the services performed by
Aristotle's quantified Logic. Perfect identity of
definition and definitum may be laid at the foundation,
and partial identities referred to it as its cases; which
is done by Professor Jevons when he uses the sign =
for the copula. For his formula " All men are mortal
men," $A = A B$, means after all just the same as the
old formula " All men are (contained under the class)
mortals," or A is B. But what I deny is, that we
should have, in such a system, an expression of the
process of thought as a subjective action; on the contrary, we should have in it *a mode* of reasoning, and
not reasoning itself, although it would be a mode
applicable to an object-matter equally wide as that
of mathematic. At any rate I should strongly doubt
the " supreme importance"[1] of teaching such a logic
as this. I should suppose rather that any one who
had once shaken off the Aristotelian quantified system,
as practically an incumbrance, either in ordinary
matters, in science, or in philosophy, would be in no
hurry to lay his neck under the yoke of any *substituted
similar;* just as I imagine that those who have once
been freed from the tyranny of Romanism would
long and seriously hesitate, before submitting themselves to the domination of a scientific Priesthood of
Humanity.

There is, besides, one important direction in which
such an artificial system as I have now described
would be of no use whatever, I mean in order to
analyse our own *meaning* in the use of words. For
such a system supposes every word used to have its
meaning fully known already, and begins only when

[1] *Principles of Science*, Vol. I. p. 129.

that meaning has been ascertained. But the process of thought subjectively is occupied at least as much with the analysis of the significance of words to us, as with the relations of the things, for which they stand, to each other. There is *meditation* in thought, as much as there is calculation in it. The system supposed provides only for the latter ; its words are calculi or counters ; their subjective analysis is taken as performed; its logic is a reasoning "from the teeth outwards."

There has always, since the days of Erigena, been a healthy feeling of the uselessness and oppressiveness of formal logic. But there was so much indisputable truth supporting it as its root, and carried upwards into its organisation, that in spite of the increasing disgust it persisted in surviving. I think I have now pointed out the exact spot upon which the cancer has fastened,—the quantification of propositions, in consequence of substituting comprehension for intension and extension. If I should prove right in this opinion, Formal Logic will soon have to be reformulated, and within much narrower limits.

The logical branch of metaphysic, however, which we may fitly call Metalogic, will have received, by the same stroke, an important development. The analysis of thought as a process will have been exhibited in its true relation to the analysis of perception. And an answer will have been given not only to the mechanical, unanalytic, unphilosophical, error of the direct consciousness, or separatist, school, which is pre-eminently English, but also to the views of an opposite school, which is pre-eminently German, that which, while seeing the true nature of the process of thought, in point of its being analytic of all existents or ready-made things, yet refuses to see the derivation of this process from perception, the dependence of the conceptual order on the perceptual, of which it is a modification. I am speaking, of course, of Hegelianism.

The great merit and achievement of Hegel, as I have remarked elsewhere also, was to show the true method of the process of thought, its movement by means of contradiction, of the combination of subject with predicate, two differents made provisionally into contradictories, in the proposition. But as Hegel wanted not only to analyse thought, but also to explain how there came to be an Universe, he had to introduce the "Real" or "Substantial" element, Aristotle's οὐσία, into his system somewhere or other. Like all genuine "Realists," he introduced it into things, not in their character of Percepts, but in their character of Concepts; and the system became in consequence a piecing together of those scattered portions of the chain of redintegration, into which conception breaks it up,[1] and that not in a perceptual order, but in a conceptual ; the fact of movement and sequence being assumed to belong to concepts in virtue of their property of Negativity; notwithstanding that movement and sequence are characteristics of presentation and spontaneous redintegration in order of perception, which is a condition upon which their modification into a conceptual order depends. It pieced them together into a chain of concepts beginning from the poorest and most general in content, and ending with the richest, for we have seen that, in categorical affirmative predication, the subject is modified by the addition of the predicate and

[1] See above, p. 301.

is enriched by a new attribute. The Universe was thus represented as a single vast *Scala Generum*, and the problem posed in Aristotle's Apodeictic was professedly solved by Hegel's Dialectic. Aristotle never could, or rather he never did, in those writings which have come down to us, give a complete List of Genera. He probably allowed himself to be too much hampered by Percepts, commonly called Facts. But when all Perception had once "gone up" into a real and concrete Concept, there was no insurmountable difficulty in rolling the conceptual snowball, until it gathered all that experience had to offer.

Hegel is in this respect the true successor of Duns Scotus. But he is more also. Inasmuch as he belongs to the modern period, the period of Reflection in philosophy, he is enabled to give an "absolute" colour to his realism, because reflection is all-embracing, being the perception of the double aspect. He is more than Platonic, he is Neo-Platonic. He is thus the successor of that earlier and perhaps greater Scotus, Scotus Erigena. For though Erigena has been fairly called a Nominalist, yet he was a Nominalist of a transcendent type. Real existence, for him, lay neither in percept nor in concept, but beyond them both: it was ὑπερούσιον, transcendent existence.[1] The method by which he found himself enabled to treat of things which surpassed all our cognitive faculties was the logical distinction with which the *De Divisione* opens, namely, between things that are and things that are not. Hegel's method is precisely similar; and therefore it is that Hegel stands as the

---

[1] De Divisione Nat. Lib. I. § 25. " Et hoc dignum quaesitu video . . . . , sed etiam intellectui esse." See also Prantl, Work cited, Vol. II. p. 20 et seqq.

representative, in modern times, of the Neo-platonists of antiquity and of the strict Platonic Realists of scholasticism.

§ 10. There remains, in order to complete the whole subject of the present Chapter, the consideration of percept and concept in the reflective mode of consciousness. The analysis which we have just made of them in the primary and direct modes is itself an exercise of reflection in both perception and conception. For without a return of consciousness upon itself we could neither have distinguished an object from our perception of it, and have thus replaced objects by percepts, nor have distinguished the trains of objects as percepts from the reasoning about them as concepts, that is, have distinguished the conceptual from the perceptual order. What we have now to do, therefore, is, by yet another return of consciousness upon itself, to say what it is that we have been doing in drawing these distinctions, to describe in general terms what reflective perception and conception consists in.

When we consider that a concept contains the moment of the onward movement in thought, that is, in redintegration accentuated by attention, while a percept in thought contains only a single member of the train which is its content, it is clear that a reflective concept must contain this moment of onward movement in trains consisting of reflective percepts. But a reflective percept, as shown in Chapter II., is nothing more or less than a perception of the double aspect, subjective and objective, in objects. The forward movement of a train of such percepts in thought is reflective conception, and gives us the conceptual order in the reflective mode. The difference between simple conception and reflective is, that the train of reflective percepts, out of which reflective conception is generated, is not spontaneous; there is no spontaneous redintegration of reflective percepts, as there was of simple percepts; and for this reason, that reflective percepts suppose a long previous train of primary conception before they are reached, and are never held fast but by an exercise of attention. Reflective perception is itself a result of reasoning applied to primary percepts.

Reflective perception is the perception of the double aspect, subjective and objective, in percepts; reflective conception is the reasoning with those percepts, is simple conception *plus* the double aspect in the percepts and concepts which are the content of the train. But simple concepts have no other second aspect, except their aspect as percepts. They give back, in consequence of attention, percepts modified. Percepts passing through reasoning come out concepts. The *objective* aspect of concepts is *mediate*, that is, belongs to them in right of their being modified *percepts*. It is the double character of its content, and not any difference in its process or law of movement, that distinguishes reflective from simple conception. The reflective mode of conception works with richer and finer materials, but does not deal with them on different principles from those of conception in the primary and direct modes. Its object-matter is double in point of aspect; that is the whole difference.

Book II.
Ch. V.

§ 10.
The Reflective
Categories or
Modals.

In reflective conception we begin with no less, and we end with no more, than the perception of the double aspect, subjective and objective, in everything. Just as the simple conceptual order began by modifying percepts into concepts, and ended by retransforming concepts into percepts again, so reflective conception begins with double percepts and ends with double percepts, being itself the process between, the comparison of concepts which can be transformed into reflective percepts at any stage of the proceeding. There is nothing in the process of reflective conception *per se* to distinguish it from the process of simple conception. Its content only, that is, the matter of its judgments and propositions, is different, being double, consisting of reflective percepts, not of percepts simply.

The subjective and objective aspects together are the proper object of reflective perception. This is not the only case of double or obverse aspects, but it is that upon which the perception that others are double depends. Double or obverse aspects are the result of perceiving the equivalence, identity, coextensiveness, or logical convertibility, of two or more different terms or objects. One instance of this we have already had in the case of percept and concept, which are obverse aspects of each other, though not subjective and objective aspects respectively. They are obverse *at the limit* where the extension of a general term becomes ideally perfect; and the perception that the perceptual and conceptual orders are obverse aspects of each other is a special case of reflection, which itself is the perception of the subjective and objective aspects. It is a consequence of this, that the concept, or extension, is sometimes con-

Book II.
Ch. V.

§ 10.
The Reflective
Categories or
Modals.

sidered as the *objective* aspect of the corresponding
percept. or intension.

It is quite possible to perceive identity or equivalence in simple conception, without reflection. Two
movements of conception. forwards and backwards,
that A is B and that B is A, put together are sufficient
for this. When the A and B are respectively feeling
and felt. or perception and thing perceived, then we
have reflective perception. or perception of the subjective and objective aspects, an identity of a particular kind. and the verification of all the rest, which
may be subsumed as cases under it. And it is probable that the circumstance of reflective perception
being a single moment with a double aspect has been
the cause of that mistake, mentioned above, of supposing that all judgments, which have convertible
terms for their object-matter, are single judgments
expressible by single propositions, as, All A is all B;
a mistake which might have been obviated by distinguishing the terms of a judgment from the movement between them.

Thus the perception of equality is not reflective
perception, but is a perception analysable by a
double movement of conception. from A to B and
from B to A. When we say " A is equal to B," we
really make two propositions, not one; and "equal
to" is really part of the predicate of the proposition. Mathematical equality again is not by any
means the same thing as logical convertibility. This
latter is not merely a *measured* equality, a sameness
in point of magnitude. but includes an equality of
kind as well: it is an equivalence between differents. A special kind of equivalence between differents, namely. a sameness in point of measured

PERCEPT AND CONCEPT. 365

Book II.
Ch. V.

§ 10.
The Reflective
Categories or
Modals.

quantity, is that which is intended by *mathematical equality.*

The perception of equality is afterwards farther explained by subsuming it as a case, falling under the more general concrete case of the perception of identity, in kind as well as in magnitude, coextensiveness of differents, or logical convertibility of differents. Equality is an identity of a particular kind, identity in respect of magnitude. But there is a more elementary identity still, and one which also precedes reflective perception. There is the identity which, in expression, is mere tautology; the identity expressed by the Postulate, A is A. There is difference even here, the difference between the times of recurrence of A. This identity, as we shall see in the next Chapter, is an essential requisite of the conceptual order, in primary as well as in reflective consciousness. It must not be confused either with concrete general identity, or with reflective identity, in perceiving the subjective and objective aspects, or with the special identity of magnitude, which is equality. It is the primary source of all. The perception of equality does not presuppose the perception of reflective identity, any more than the perception of identity in point of kind presupposes it. On the contrary, these two identities are both pre-requisites of reflective identity, identity of perception and thing perceived. We perceive equality, and sameness of kind or quality, without perceiving that they are cases of identity in general. But probably the first perception of total identity of differents is the identity of reflective perception, the identity of the subjective and objective aspects, at the moment when reflection arises out of primary consciousness.

Book II.
Ch. V.

§ 10.
The Reflective
Categories or
Modal.

We can now see the advantage derived from analysing redintegration into the processes of perceiving and conceiving; which advantage consists in this, that we can break up reflective perception itself into its component movements. And the analysis of consciousness into its modes, primary, reflective, and direct, is in its turn controlled by that into percept and concept, just as, in another respect, we have seen this latter analysis controlled by that into the three modes.

The question next arises. In what shape does conception give us back the reflective percepts? Simple conception gave us back simple percepts in the shape of concepts, definitions, and the generic scale. What corresponds to this in the case of the reflective percepts? Conception is characterisation. What character do the reflective percepts bear, when they are conceived or treated as concepts? It must be such a character as to leave them still reflective, that is, having still their double aspect, subjective and objective. At the same time it must be a character not drawn from without, from any mode of consciousness beyond or above reflection, for we know of no such mode. In other words, it cannot be a characterisation of the whole order of reflective percepts, but of some parts or features of it compared with others; a distinction of classes of reflective percepts.

There is only one feature in reflective percepts which satisfies both these conditions, and which can therefore serve as the point of distinction and comparison. It is the feature of *Time*. Time belongs both to the objective and the subjective aspect of all percepts; time is within reflection, and does not be-

BOOK II.
CH. V.

—

§ 10.
The Reflective
Categories or
Modals.

long to any (supposed) mode of consciousness beyond it. But also it is evident that no ready made, and therefore, for the present purpose, arbitrary, distinctions in time can be the basis of comparison. It is time itself in its utmost generality. Such a distinction, for instance, as time past, present, and future, which is a merely popular and empirical distinction, taken alone, would be quite irrelevant to the purpose.

We want to characterise different modes of existence, or of perception, by their time relations. And we have, to start with, the general perception that all of them must have the double aspect, subjective and objective. Conception, however, is reasoning; we want to characterise in order to attain real knowledge; we start then with that kind of percept which is immediately perceived, with percepts such as we have them in *presentations*. We generalise presentations. All true knowledge, all real existence, must be capable of being the object of presentative perception to some consciousness or other. This is the notion of *Real Existents*, or (subjectively) *Actual Percepts*.

But what of the content of the rest of Time, other than that of any given presentation? Two things are certain with regard to it, first, that any particular portion of it, as compared with actual percepts, is uncertain; second, that, if any portion is real, it will on appearing in actual consciousness have such features as all actual percepts have, duration being one, whatever the rest may turn out to be. The first consideration gives us the characterisation of *Contingently* existing things, or (subjectively) *Possible Percepts*. The second gives us that of *Universally* existing things, or (subjectively) *Necessary Percepts*.

Book II.
Ch. V.
──
§ 10.
The Reflective
Categories or
Modals.

Whatever may or may not be, according to our present knowledge. is, when conceived as an object so far as it is known, contingent. Whatever must be perceived or thought, no matter what else is perceived or thought along with it, is, when conceived as existing, that is. in its objective aspect, universal, omnipresent.

We have accordingly what are called the Modal Categories. which I have attempted to show are Categories of Reflection, thus accounting for their double character. We may arrange them in a tabular form, and in generalised terms, thus:

The Modal Categories or Categories of Reflection.

SUBJECTIVE ASPECT.	OBJECTIVE ASPECT.
Actual	Existent.
Possible	Contingent.
Necessary	Universal.

It is a list of the most general classes of reflective concepts; and it has the advantage of exhibiting the derivation of the Modals, as they are called, from the reflective mode of consciousness, as well as that of keeping them clear of the confusing admixture of their own contraries. Modals, it is true, like every other kind of concepts, have their contraries; but why mix up these with the question of their nature? It is common to all contraries to be formed by means of the process of contradiction, as I hope to show in the next Chapter; and a concept is never fully understood until it is contrasted with its contrary; and therefore the present list is by no means final. But why Modals alone should never be exhibited except

Book II.
Ch. V.

§ 10.
The Reflective
Categories or
Modals.

as part of a table of contraries I have never yet been
able to discover, unless it be a blind following of
Aristotle's example in his *De Interpretatione*; where,
however, this arrangement is quite in place, being
dependent on the analysis of logical propositions;
—combined always with an insufficient analysis of
the phenomena of reflection.

The modal or reflective categories are the highest,
or most general, notions which we can form of the
universe at large; and the importance which they
derive from their reflective origin will be recognised,
if we remember what the scope of reflective consciousness is, compared with that of direct, of philosophy as compared with science. It is this, that
they help to fill up the empty *cadres* of reflection,
where empirical knowledge stops. In the constructive branch of philosophy, they afford a framework
upon which any notions or imaginations, which we
may indulge, must be stretched. They fill infinity,
while they penetrate and pervade our present actually
phenomenal world; which stands to them, as we
cannot but think, in the relation of instance to rule,
of case to general law.

For it should not pass unobserved, that the list of
modals now given does not give them as contraries
in any way. It does not disentangle them, each
from its own contrary, only to bring them into the
position of contraries to each other. There is no
antagonism between them. It is the conceptual point
of view, the reasoning for the purpose of knowledge,
by and from which they are distinguished. They are
no more fixed determinations of supposed absolute
things, than are the *genus*, *differentia*, and *species* of
simple conception. They are mobilised forms of

Book II.
Ch. V.,
———
§ 10.
The Reflective
Categories or
Modals.

thought for laying hold of the universe of reflection, as simple concepts lay hold of the world of primary and direct perception. And they are as completely and as inevitably derived from reflective, as the simple categories are from primary. perception.

I will give an instance for the sake of obtaining definiteness for my meaning, at the cost of a little anticipation. Take the question of Freedom of Will in choosing. Freedom is a phenomenon belonging to reflection; the perception of being free to choose, or to will, arises only in and with the perception of self. It is a case under the general notion of Possibility; we *can* act or not act in a particular way. We thus have to consider freedom in conjunction with the other two categories of Actuality, or Real Existence, and of Necessity or Universality. And the question is—*What* freedom is in analysis or connotation, what *more* it is than that perception of being free, which is a phrase denoting what we intend to speak of.

This question is soluble by distinguishing, by a further modification of the distinction drawn above in this Chapter, two ways of regarding the course of time with its moving content of conscious states;— 1st, transversely or statically; 2nd, longitudinally or dynamically, from the present moment forwards into the future. In the first, or transverse, way of regarding the course of time, everything is necessary, a case of universally applicable laws or general facts. It is from the second point of view that things appear *contingent*, and our own action as *free*. But if we consider in the next place, that this appearance must be capable of being harmonised with the first appearance, and consider also the fragmentary and partial

PERCEPT AND CONCEPT. 371

Book II.
Ch. V.
—
§ 10.
The Reflective
Categories or
Modals.

character of the view obtained by the second method,
(which it is, because, while immense tracts can be
embraced by the first, only the next moment or two
can be seen by the second),—we are led to conclude
that the Truth of things is given rather in the transverse or statical view, and that consequently universal
law or necessity is the truth of things and of actions.
We can then assign its true meaning to the sense of
freedom, as well as of contingency; namely, that
what we choose now is not *cause* but *evidence* of what
the end will be to which it leads, or the whole to
which it belongs; and that like the contingent event,
so also the contingent act, or choice, is merely the
coming into consciousness of the pre-existing order
which was before hidden from us. We *are* part of
that order; our choice *is* its actuality. In short we
may regard the whole Universe, statically, as an
eternal Now; or we may regard it, dynamically, as
an eternal nexus of conditions and consequents; or
we may vary slightly these methods and regard, first,
the course of time at a single moment transversely,
which gives us that moment as *being;* and secondly,
longitudinally, which gives us that moment as
action.

I will give one other instance in illustration of
the bearing of reflection generally upon the constructive branch of philosophy. Man, in the infancy of
the race, began by looking at all nature, all objects,
as personal and conscious, like himself. He has
ended, *up to the present time,* by eliminating personality and consciousness from everything but God
(*i.e.* the supposed Unknown or Beyond of experience) and himself. *Will* he proceed to eliminate it
from God? *Can* he eliminate it from himself?

Book II.
Ch. V.
———
§ 10.
The Reflective
Categories or
Modals.

Clearly not the latter; there *must be* a subjective aspect. But if a subjective aspect *at all*, then necessarily of *everything*. Personality and consciousness return into all nature again; but they return in the shape of a characterisation, a determination of reflection, not of direct perception, not in the shape of an immaterial substance or agent, nor as *seated* in any other centre than man.

So far is the result of reflective analysis. But when we go on to the questions of history and of origin, and ask *Whence*, then, has man his consciousness, *how came* he to be its seat? the answer must be (in some form or other), From that Nature of which he is a part, and which is *in part* the object of his direct, as well as of his reflective, perception. Man and Nature, then, are of a piece, and man is the seat of consciousness. Inevitably the question arises,—Is not nature so also, though man may have no direct perception of the fact? This is the main question of the Constructive Branch of Philosophy.

CHAPTER VI.

CONTRADICTION AND CONTRARIETY.

Book II.
Ch. VI.

§ 1.
The Principle
of Contradic-
tion.

§ 1. It is far, nay incomparably, more needful at the present time to work at the dry detail of metaphysical analysis, the foundations of analytical philosophy, than to indulge in speculations, however interesting their questions or dazzling their replies, in the domain of constructive philosophy. Metaphysic is a pressing need. Why are mankind everywhere at war with one another on questions of religion? Not because there are not religious men on this side or on that; not because dogmas, being indifferent (which they are not), are taken as essential; not because religion is mixed up with motives of personal and party interest, vanity, and ambition; but mainly and fundamentally because religious men themselves are ignorant of the analysis of human nature, including the mode in which religion belongs to it, and the relation which it bears to its brother elements in it. Until we have an accepted metaphysic, religious concord is impossible.

Again, what are the internal causes which chiefly hasten, even if they do not alone produce, the decay, old age, and death, of nations which have once

BOOK II.
CH. VI.

§ 1.
The Principle
of Contradic-
tion.

enjoyed a vigorous prime? There comes a time in
the life of prosperous nations, when the mass either
of the whole population or of the governing classes
find themselves in a condition of ease and affluence,
and turn their thoughts to individual enjoyment.
Their life of hardship and effort, the direction of
which was determined by the necessity of struggling
with some external force, is changed into a life the
direction of which they have to choose for themselves.
I speak of the individuals composing the mass either
of the whole population or of the governing classes,
who give its stamp to the whole tone of society, and
direct the activity of the whole. Every such period
in a nation's life is a period of crisis. A Choice of
Hercules has to be made. If such a period coincides
with one in which old religious dogmas are falling
away, and religious sects are falling asunder, the crisis
is doubly critical; for a main circumstance that would
otherwise have guided the choice is absent. If self-
knowledge is important to an individual in every case
where he has to make his own choice in direction of
life. an agreement on the true principles of conduct,
founded on agreement on the true analysis of human
nature, is no less important to a nation. Not that this
supplies the place of religion, but that it enables
religion to be generally efficacious, when otherwise it
would be paralysed through religious dissensions. I
think I do not overrate the importance of the services
which a true metaphysical analysis might render,
though mediately and remotely. In medicine, a
sound diet is not to be despised, merely because it is
not a panacea.

Taking leave, then, of speculations like the last of
the two brought as illustrations at the conclusion

of the foregoing Chapter, I return to the main course
of the series of distinctions which are the purpose of
the whole. We may, in accordance with the proof
given in the foregoing Chapter, define reasoning, or
the conceptual order, as *the forward moving sequence
of conscious states in redintegration, subject to the law of
parcimony*, that is, modified by the volition of reducing them to the order which is the simplest in
form, and at the same time the richest in content.

In this sequence, we have at the outset to distinguish the time which belongs to the sequence
itself from the times, or from the times and spaces
together, which are occupied by the images which
compose the sequence. These may sink into points
or swell into infinity; at the same time they are strung,
so to speak, on a line of time, by the mere fact that
they follow one another in thought. I may think
now of the earth's orbit, next of a bee's wing, next of
the beauty of a just action. All these images become
concepts in thought. Their relations to other images,
and the relations of their constituent parts to them
and to one another, are what we mean to express by
ranging them in the generic scale. The generic scale
and all its parts is a consequence not a condition of
predication, as was shown in the last Chapter; it is so
because its formative general categories, the categories
of logic, *genus*, *differentia*, and *species*, are themselves
formed by attention, comparison, and judgment. These
logical categories are necessary to thought; to attend
to and compare percepts is to turn them into concepts,
and to subject them to a generic scale. But the
genera, *differentiae*, and *species*, which are at any time
the *content* of this scale, are changeable in the highest
degree; not being, as they stand, the order of nature,

Book II.
Ch. VI.

§ 1.
The Principle
of Contradic-
tion.

Book II.
Ch. VI.

§ 1.
The Principle
of Contradic-
tion.

but depending on the degree of our insight and knowledge from time to time. The logical categories are mobile in applicability, fixed in their own nature; indifferent to their content, necessary in their form.

It is to the line of time upon which these sequences of concepts are strung that the point of division and connection belongs, which in propositions is marked by the copula, the point fixed by attending to some feature in the redintegration. Without this point there would be no concepts, no comparison, no judgment, no predication, no thought, no reasoning. The first step in making a comparison or determining a relation is to form a concept, to fix on what you are going to compare and to distinguish it for that purpose from everything else. This is the work of attention. You call it, suppose, A. Then A, because it is expectant, is a concept. And by A you mean A, and nothing else. But you proceed to fulfil the purpose for which you fixed upon it, and to compare it or bring it into relation with something else. This *something else* is something or other offered by spontaneous redintegration, which has been previously separated from A by attention. A which was, in attention, *not* not-A, now, in predication, *is* not-A, namely (suppose) B. The proposition expressing the judgment, A is B, arises by overleaping the point of Contradiction, the point fixed by attention to distinguish A from not-A.

The facts which have been just described make up what is known as the Principle of Contradiction. The principle of contradiction is a short-hand expression of these facts, of the nature and significance in logic of attention and volition in reasoning. The principle of contradiction is thus the ultimate basis of

Book II.
Ch. VI.
———
§ 1.
The Principle
of Contradic-
tion.

logic; it is so, because attention in redintegration is so. This latter fact is the *psychological* basis of the logical principle. The principle itself has moreover been formulated in logical terms, and expressed as consisting in the Three Postulates of Logic; Postulates not Axioms, because they are volitional in origin, and define a method, not an already existing object-matter. They may be formulated as follows:

The Postulates of Logic.

1. Identity.	2. Contradiction.	3. Excluded Middle.
e.g.	*e.g.*	*e.g.*
A is A	A isn't not-A	Everything is
	or	either A or else
	No A is not-A.	not-A.

Incapable of existing without content, but at the same time completely indifferent to any particular content, these postulates, expressing the principle of contradiction, are the leverage of conative consciousness, the point of resistance in the machinery by which it works.

The consequences which flow from this principle, the modifications and distinctions which arise from it, and not from the particular kind of content in or upon which it works, these are the object-matter of Logic as a science, the machinery of Logic as an organon. These and these alone constitute formal logic. They are established and justified, as in the present Chapter, by Metalogic, a branch of metaphysic or analytic philosophy. They will be deduced from, or shown to be consequences of, the principle of contradiction. This and the postulates which formulate it cannot be demonstrated, but must be and are assumed as self-evident. They can be pointed out and

Book II.
Ch. VI.
——
§ 1.
The Principle
of Contradic-
tion.

analysed as facts, facts universally present in a particular kind of action, namely, reasoning, and therefore necessary to that kind of action. As facts, they
are perceived as inherent in reasoning; as truths,
they are assumed as self-evident and incapable of
demonstration. You must either reason by their
means or not at all. And the truth of the whole content of formal logic depends on the truth of this
assumption.

It is the indifference of formal, or as it is often
called general, logic to *every* kind of content separately, though not to content at large, which makes
it more abstract and general than any other science.
It stands above the mathematical sciences in this respect. Its object-matter is the formal element of all
consciousness, time, divided *only* by the point of
contradiction, which is marked by the copula. It is
clearly a pre-requisite of Geometry, and not only so,
but the object-matter of geometry is far more special
and concrete; it is the configuration of space, space
already divided by difference of feeling, magnitude
of a particular kind. Logic is also a pre-requisite of
Arithmetic, for it is prior to Number. You cannot
have number till you have divided or distinguished
an A from everything else, say from B. When you
have distinguished A from B, you can call A *one*, and
B *two*. But the act of distinguishing them precedes.
Time must be distinguished into moments by attention before those moments are marked as *first*, *second*,
third, and so on. The act of attention may give us
the moments as former and latter, preceding and
following, but mere sequence in time is not number.
It is a condition of it, a step towards it. In order to
have number, all other features of the sequence, ex-

Book II.
Ch. VI.
—
§ 1.
The Principle
of Contradic-
tion.

cept that of times of occurrence, must be consciously
excluded. Number is simply and merely the *discrete-*
ness of quantity, *i.e.*, multitude. Algebra, again,
though in point of method more general than Arithmetic, is not more abstract in respect of its constituent elements. Its methods enable you to treat of
numbers, and affections of number, under general
denominations, that is, symbols, which may or may
not prove to have actual numbers corresponding to
them; but the full notion of discrete quantity, multitude, and measure of multitude, is always involved.
It is not more but less general than arithmetic, in
respect of complexity of content. It *adds* determinations of number to those already employed in arithmetic.

Our starting-point, then, is the Principle of Contradiction, formulated as the Postulates of Logic; and
their pre-supposition, the object-matter into which
they are introduced, is the train of *differents* in redintegrations in perceptual order, or the sequence of
different images composing perceptual redintegrations. Observe that in these redintegrations we have
only *differents;* the images are different from one
another, they are not contradictories, nor contraries,
nor privatives, nor negatives, nor relatives. It may
be said perhaps that they are relatives, since they are
differents; difference being a certain sort of relation.
But this apparent inconsistency arises from our being
compelled to use general terms in speaking of everything, even of things in their first intention. I mean,
we are compelled to call different things *different*, even
when we speak of them before they have been *thought*
as different. We often remember that, in dreams,
different images have replaced one another without

Book II.
Ch. VI.
————
§ 1.
The Principle
of Contradic-
tion.

our having been conscious of the moment of change from one to the other; the reason is, that in these dreams we are not reasoning but merely redintegrating, though possibly redintegrating the *results* of reasoning; the moment of volition is wanting in the connection of the different images. It is a difference of this kind that is involved in the perceptual trains of redintegration before the introduction of the point of contradiction. To think of two things as different is not merely to perceive difference; it is something more; it is to bring them under a relation in thought. But relation in thought depends upon the introduction of the point of contradiction. Without this, there is no connection in *thought*, there is only connection in *fact*, between them.

When we have formed such categories as different, relative, contrary, and so on, we apply them to everything, even to the perceptions which are their prerequisites. The very universality of their applicability, which is their glory and the justification of thought as an exhaustive process of reaching truth, is a misleading circumstance if we take these categories as ultimate and inexplicable, and do not submit them to a careful analysis. And language itself, consisting as it does of terms already generalised, tends to blind us to the difference between first and second intentions, to the double source of our general terms.

That source of them which immediately concerns us is the point of contradiction, and its introduction into trains of differents. We fix upon one of these differents by attention, A, and the first thing that arises is a single proposition, a *negative* one:

A isn't not-A.

Book II.
Ch. VI.

§ 1.
The Principle
of Contradic-
tion.

This gives us the contradictory terms, A and not-A, by virtue of the postulate of Identity, A is A; for we have A is A, and not-A is not-A; and by the third postulate, there is nothing that is not either A or else not-A. The three postulates hang together. State one and you state all; contradict one and you contradict all. But it is from the second postulate, the postulate of contradiction, that contradictory terms arise. In order to explain what contradictory terms are, we have to go to the postulate of contradiction.

In the next place it is to be observed of these contradictory terms, that the negative one, not-A, has no content. There might have been *in its stead* a term with a positive content, and then the proposition might have been an affirmative one, A is B. But as it is, in the proposition, A isn't not-A, the term not-A has literally no content. All its possible content is in its contradictory, A. But A is excluded from it *in toto*. This is only to say, in other words, that there are no such things as *wholly* contradictory percepts, or that contradictory terms, as such, are solely categories of thought, not of perception; for though A may exist, and indeed is assumed to exist, and have a positive content, yet it does not exist *as a contradictory* unless its contradictory also exists; and this it does not, not-A having no content. Of two wholly contradictory terms, the one is thought as existent, the other as non-existent. The reason of the restriction, *wholly*, will be seen when we come to treat of Contraries.

In the next place it is to be noted, that the positive content is in the subject and not in the predicate of the proposition, A isn't not-A. A is a percept, an

Book II.
Ch. VI.

§ 1.
The Principle
of Contradic-
tion.

existent, what in Grammar is a substantive. The contradictory term to it, not-A, is a non-existent, is not a percept, having no content. To state the existence of not-A, as a contradictory term to A, you must employ the contradictory proposition to A isn't not-A, *viz.:* A is not-A, which exhibits the absurdity. Or if by an Hegelian turn of expression[1] we adopt the converse of the proposition, and say Not-A is A, this can only be a true proposition in the sense that A is an element in the *logical* term not-A, while the appearance of existence, of perceivability, in not-A is given to it solely by its being put in the place of subject of the proposition, *as if we knew it independently of its predicate.* This proceeding contains the substance of Hegel's method. The negative member of a pair of contradictory terms, which is a pure creature of logical method, analogous to imaginary quantities in mathematic, is treated by Hegel as if it were a concept with a perceptual content. The "*Nichts*" at the beginning of the *Logik* is the first instance of it. But this can delude us only so long as we are unaware of the double source of terms of thought, the perceptual and the conceptual, and of the mode in which they are combined in reasoning.

Two things, then, derive immediately from the postulates, contradictory *terms,* A and not-A, and contradictory *propositions,* A isn't not-A, and A is not-A. And just as there is no perceptual *content* in the negative contradictory term, not-A, so also there is no *truth* in the contradictory proposition which affirms that negative contradictory term of its own contradictory; there is no truth in it because it con-

---

[1] *E.g.* "*Nicht ein Dieses* ist ein Allgemeines der Reflexion." Subj. Logik. Das Singulare Urtheil. Werke, Vol. V. p 91.

CONTRADICTION AND CONTRARIETY. 383

Book II.
Ch. VI.

§ 1.
The Principle
of Contradic-
tion.

tradicts the postulates. The principle of contradiction is the logical source of truth.

I subscribe, then, but with the reservation expressed above by the word *wholly*, to Herr Pastor
Knauer, when he says in his most philosophical little
treatise *Conträr und Contradictorisch*,[1] that there are
no such things as contradictory concepts, but only
contradictory judgments. "Contradictorisch sind
auf assertorischem Boden nur die Behauptungen von
Existenz und von Nichtexistenz." And again, "nur
wenn ein und derselbe Begriff ab- und zugesprochen
wird, findet wirklich die Contradiction statt." To
say that there are no wholly contradictory concepts
but only judgments, and to say that the term contradictory *in toto* of another term having a positive content has itself no positive content, are one and the
same thing. I think nevertheless that it is rather a
roundabout way of arriving at this, to deduce it from
the Kantian table of Categories, by referring it solely
to the moment of Modality. The analysis of the process of judgment seems to me a far more direct and
conclusive way.

Knauer's criticism of Hegel's views on contrary
and contradictory concepts is also excellent.[2] Hegel
admits the existence of contradictory concepts, but
says that they, as well as contrary concepts, are
determined in a superficial manner by ordinary logicians. His own view is, that they are moments in
the evolution of the Concept; belonging to that stage
of it when it appears as the General Concept (*allgemeiner
Begriff*), and expressly visible in the particular form
of Judgment which exhibits the general concept,

[1] Halle, 1868, pp. 86. 87. 88.
[2] Work cited, pp. 119-121.

Book II.
Ch. VI.

§ 1.
The Principle
of Contradic-
tion.

namely, the Disjunctive Judgment. The disjunctive judgment, with Hegel, is the exhibition of what we commonly call a genus and its several species. And *contradictory* is the term applicable to the several species as they are exclusive of one another; *contrary* the term applicable to the same species as they are identified with one another in the genus.[1]

If it should excite surprise that this is all that Hegel has to say upon so fundamental a point in philosophy and logic as the nature of contradictory and contrary opposition, the explanation is simple. It was precisely the point which Hegel's theory made of no importance. It is a pre-supposition of his theory that this point is not fundamental. The evolution of the concrete concept is his fundamental idea; it evolves itself by *Entgegensetzung*, a concrete opposition containing undistinguished the purely logical opposition of contradiction, and the opposition of content which is contrariety. The former opposition gives the motive power, the latter the order and arrangement, of the evolution. Thus the pure Nothing, *Nichts*, at the beginning is logically opposed to the pure Being, *Sein;* hence the *movement* between them. There is no opposition of content, no difference of content at all, between them, until they are conceived *together;* then they are perceived to be different in *content*, but at the same time to be a process, a *Werden*, not (either of them) a state or thing. The whole makes one indistinguishable process of opposition or becoming. *Entgegensetzung* or *Werden*. To analyse this process, to show what is due to perception, what to conception, what part of

[1] Hegel, Subjective Logik. Werke, Vol. V. pp. 54-55 and p. 101.

Book II.
Ch. VI.

§ 1.
The Principle
of Contradic-
tion.

the opposition is due to content, and what to logical contradiction, would be to destroy it as a theory of the universe. It supersedes the principle of analysis by the principle of concretion. The Hegelian logic is *radically* different from the Aristotelian, the Aristotelian resting on an analysis which the Hegelian obscures and obliterates.

§ 2. It is next in order to follow up the question of contradictory terms, and to see how we pass over the logical point of contradiction in propositions, and apply the method of contradiction in positive predication, *without* contradicting the postulates by which we move. We have A distinguished from not-A as its contradictory term. How get beyond the apparently closed circle of the postulate of Identity, A is A? It is experience, brought again before us in perceptual redintegration, that decides the question. A is a concept, an expectant; but apparently expectant of nothing, or with its expectations deceived, since by the postulate of identity A is nothing but A. But everything depends upon *what we mean* by A, what its perceptual content is. A, to use Aristotle's phrase, σημαίνει τι, and the question is, τί σημαίνει ;

Suppose A to be a singular existent, since it has been already shown that percepts are treated as substantives by being used as subjects in propositions; suppose it is the Hat on the table before me. We have it in a proposition thus:

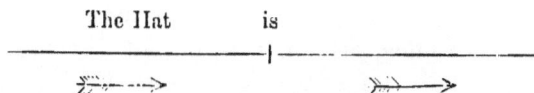

The Hat   is

Aye, *What?* It is well to leave a blank. It is plain that it can be nothing but *The Hat.* Yes. But it may be, nay must be, aye *is*—The whole Hat. Precisely so, the hat, the whole hat, and nothing but the hat. We have before us a content of perception, and the next moment of the proposition, its predicate, consists of that content or part of it. The hat has qualities in it, and relations between these, and relations to other objects outside itself, and to various percipients, and to various times and places. It is a portion of the perceptual universe in connection with other portions. These qualities and relations are parts of the whole hat, *determinations* of the hat to be *in every particular what it is;* and they furnish the predicates of the hat as subject. The copula, *is,* expresses the coalescence, that is, the uniting into one image, of the predicates with one another and with the hat. Our *perception* of what the hat is and what its relations are, as percepts, determines our choice of the predicates; is their *conditio existendi* as predicates; while conversely their application as predicates is the *conditio cognoscendi* in our logical or reasoning cognition of the hat. The subject determines its predicates in perceptual order, in virtue of its being a percept; the predicates determine the subject in conceptual order, in virtue of their being concepts. The determinations too, which are expressed as predicates, are all adjectival determinations in the first instance, whatever tricks we may play with them afterwards. I mean that, as given by experience, they are adjectival. In saying, *e.g.* the hat is a cover for the head, we do not mean, at least if we merely repeat experience, 'the hat is one of the class called covers for the head.' We mean 'the hat stands in a

CONTRADICTION AND CONTRARIETY.   387

Book II.
Ch. VI.

§ 2.
Contradiction
in Predication.

certain relative position to the head.' The imagery
before us consists of a hat and a head, and not of all
sorts of covers for heads. Yet with all this fulness
of predication we have not contradicted the postulates; we have merely determined their *content* in a
particular case. All these predicates are not not-A
but A. They never were not-A; they were not
thought at all; they lay in the indetermination of
experience either newly given or registered in
memory. They come from perception and are
ranged and ordered by conception; they are not
creations of The Concept.

§ 3. We have now to follow up the principle of
contradiction in its application to concrete predication, in such propositions as those just described.
In the first place, what is the circumstance which
differentiates predication of this sort from the abstract propositions in which the postulates are expressed? It is this. In the postulates, the contradictory or identical terms appear as subject and
predicate of the same proposition. Those terms
are A and not-A, and the predicate not-A is the
contradictory of the subject A. But the concrete
proposition, A is B, is a determination of the subject
by the predicate. Its contradiction therefore must
be a contradiction of that determination. The thing
affirmed by the original proposition must be denied
by its contradictory proposition. Again, since it is
the characteristic of the negative member of a pair
of contradictory propositions (just as in the case of
contradictory terms) to have no perceptual content,

Book II.
Ch. VI.
——
§ 3.
Contradictory
and Contrary
propositions.

as in *A isn't A*, which is the contradictory of *A is A*, it follows that the concrete contradictory proposition, of which we now speak, must *affirm nothing* in denying its original, affirmative, contradictory. It must *not determine* in any way the subject common to both.

Accordingly we have, as concrete contradictory propositions, such propositions as:

This black hat is heavy,—
This black hat isn't heavy;

where nothing whatever is affirmed of 'this black hat,' the common subject, by the negative member of the pair of propositions. The determination 'heavy' is denied of it, and nothing more. If however I were to take the contradictory *term* to the predicate 'heavy,' namely, ' not-heavy,' and affirm this term of the subject, as in 'This black hat is not-heavy,' I should get, not the contradictory of the original proposition, but the affirmation of *another* determination, opposite and exclusive of the former. I should affirm that 'this black hat' had a determination, and that this determination was the exclusive opposite of 'heavy.' I should get a *contrasted*, or possibly a *contrary*, but not a purely contradictory determination.

By contradictory terms, then, I mean contradictories *in toto*, contradictories as existents, as substantive percepts one of which exists and the other exists not. And it makes no difference what the content is, so long as that content is taken as an existent. Its non-existence, that is, its non-existence as a percept, is stated or expressed by the negative contradictory term: A——not-A; This black and heavy hat——not this black and heavy hat.

Book II.
Ch. VI.

§ 3.
Contradictory
and Contrary
propositions.

But take the contradictory terms as determinations of an already existing percept, as adjectives of a substantive, and you make those contradictory terms into *contrary* terms. Contrary terms are contradictory determinations of a subject, or existent, supposed to be given and admitted. This is their *differentia* from contradictory terms, which may be called their *genus*.

Briefly, then, we may define contradictory categorical propositions as those in which, subject and predicate being the same, the copula is affirmative in one and negative in the other. Contrary categorical propositions may likewise be defined as those in which, the copula remaining affirmative, the predicates are contradictory determinations of the subject. And the practical criterion is, that the negation in contradictory propositions lies in the copula, whereas in contrary propositions it is attached to the predicate. For instance:

This black hat is heavy. } Contradictories. Nothing is *affirmed*
This black hat isn't heavy. } by the negative proposition.

This black hat is heavy. } Contraries. The subject is affirmed
This black hat is not-heavy. } to be *something* but not heavy.

The foregoing account of this matter differs widely not only from Hegel's but also from that usually given, and rests upon the principle of concepts being the units of thought, and upon their being treated as units in the real and actually employed process of thinking and reasoning. The account usually given rests, on the other hand, on the quantification of propositions as universal and particular, and their classification under the well

Book II.
Ch. VI.
—
§ 3.
Contradictory
and Contrary
propositions.

known logical symbols A. E. I. O. Contradictory propositions are, according to this, those which differ both in quantity and quality, contrary and sub-contrary those which differ only in quality; the quality of propositions meaning their affirmative or negative character, and quantity meaning the "all" or "some" of the collective terms which are their subjects. Propositions where the negation attaches to the predicate, as in 'the hat is not-heavy,' are not provided for by this scheme, and are then classed apart as expressing what are called "infinite" or "unlimited" judgments,—a class at once useless and confusing.

This account of contrary and contradictory propositions recognises to some extent the true principle, that a negative contradictory proposition makes no affirmation of its own, but confines itself to denying its affirmative contradictory. But the controversial, discussional, origin and purpose of the formal logic, to which this account belongs, are betrayed by the limited use made of this principle. A and O are contradictory propositions, — say, 'All swans are white,' and 'Some swans are not white.' This negative contradictory. 'Some swans are not white,' is not the contradictory of 'All swans are white,' except inasmuch as the fact which it states enables us to deny it. It is not the contradictory, but *a premiss* enabling us to deny 'all swans are white;' some one has seen swans which are not white, and *therefore* 'all swans are white' is untrue. Such an account of contradictories and contraries leaves the nature of contradiction itself entirely unexplained.

Contrary propositions depend on concrete predication determining a subject by predicates; and just as contrary terms come from contradictory terms by

Book II.
Ch. VI.

§ 3.
Contradictory
and Contrary
propositions.

taking the latter as determinations instead of as substantives, so contrary propositions come from contradictory propositions, not by a change in the copula, but by a contradiction within the predicate or determination of the subject. This distinction of pairs of propositions into contradictories and contraries exhausts the division of propositions in this direction and in their character of propositions. The essence of a proposition lies in the copula, the *nexus* of judgment. When the ground of division is removed into the predicate, all further division of propositions is not a division of them in their essence but in their accidents. We have propositions divided, first, in respect of their essence, their copula. This gives us the two classes of contradictory and contrary propositions, contradictory where the copula is different, and the propositions respectively affirm and deny an existent as substantive; and contrary, where the copula is the same, a determination is expressed in both cases, and the determination affirmed by the one is denied by the other.

§ 4. It remains to follow up the division of propositions into the distinctions offered by the mode of determination of the predicate; the *logical* distinctions be it observed, which depend, not solely on the perceptual differences observable in the object-matter, but on these under the action of the principle of contradiction, or in their character of contradictory determinations. The heads under which these differences will be found to fall are the true logical machinery in actual employment and of practical use.

Book II.
Ch. VI.
———
§ 4.
Contraries.
Privatives, and
Relatives.

More will have to be said, farther on, concerning the principle and method by which this machinery itself is created. But for convenience of exposition I postpone this, until the Modal Categories have been examined.

The ascertainment of these distinctions is in one respect more simple than the part of our task which now lies behind us, the discrimination of the properties of terms from the properties of propositions. We have no longer to drive four-in-hand, so to speak: contradictory terms and contradictory propositions together; we are reduced to a pair, contradictory terms in their character of determinations of given existents, *i.e. Contrary Terms.*

At the same time, this division of the subject is not without obscurity, which has indeed lasted from the days of Aristotle. Here, as so often elsewhere, Aristotle laid the foundations. But his special treatise on the subject, his ἐκλογή or διαίρεσις τῶν ἐναντίων has not come down to us.[1] Still a great deal, and that bearing on the logic of the case, is found in what we have; more particularly in the First Book of the Physica Auscultatio, and in Books Γ and I of the Metaphysica.

This division of the subject, besides, corresponds to what Kant called *Construction of Concepts,* a division which he excluded under that title from Metaphysic altogether, and treated as a mere application of concepts to intuition;[2] and which it was Maimon's

---

[1] See Bonitz; Comm. on the Metaphysica, Book Γ. p. 1004 a. 2.

[2] Preface to 2nd edit. of the Critic of Pure Reason, p. 18. Hartenstein, 1853. See also *Fortschritte der Metaphysik,* Werke, Vol. I. p. 490, ed Ros. u. Sch. *Prolegomena &c.* Vol. III. p. 23. 36. *Logik,* Vol. III. p. 184-5, same edition.

Book II.
Ch. VI.

§ 4.
Contraries,
Privatives, and
Relatives.

achievement to bring back into metaphysic again, by means of his doctrine of Determination (*Bestimmbarkeit*).[1]

Contradictory determinations of given existents, *i.e.*, contraries,—how can they exist? In the first place it is clear that they cannot exist in the same existent, at the same time, and in the same respect.[2] For the principle of contradiction forbids it. Let us take separately these three well known requisites of contradictory determinations, and first the requisite of being predicated *in the same respect*. This requisite of contradictory determinations forces us back upon experience, upon the perceptual redintegrations. It compels us to give an analysed content to the objects or images thought under the distinction of ' this' and ' not-this.' For instance, in ' This black hat is heavy,' and ' This black hat is not-heavy,' ' heavy' and ' not-heavy' are not contraries if one is said in respect of some standard of heaviness in things possessing weight, and the other in respect of heaviness altogether, of ponderables in contrast to imponderables. The filling up, or the content, of the distinction drawn by contraries comes from the perceptual redintegrations, the logical analysis of that content is performed by drawing the distinction itself, whereby the perceptual classification of predicable determinations is transformed into a conceptual classification of contraries in various respects.

We have before us, then, the *Differents* of perceptual redintegration. The first step towards establishing a pair of contraries is to *think* a difference

[1] *Versuch einer neuen Logik. Grundsätze des reellen Denkens,* pp. 184 to 190. 2nd edit. 1798.

[2] *Dici de eodem, eodem tempore, et secundum idem.*

Book II.
Ch. VI.

§ 4.
Contraries,
Privatives, and
Relatives.

between two differents of perception by supplying in thought the *respect* or feature in which they differ. This is what we have already described. For we make a species by combining a positive differentia with a genus. But different species are not therefore contraries. A species is constituted by its own differentia, irrespective of other species, notwithstanding that all have a perceptual content. Two positive species have each its positive differentia; but for two species to be contraries, their differentiæ must be contradictory as well as positive, or in other words. the two species must be contradictory in respect of one and the same feature of the genus. And we have no longer merely to predicate a single characteristic of an existent; but we have to predicate one of two contradictory characteristics, both of which are *positive*. The *not* of contradiction falls within the predicate itself; the respect in which one different differs from another divides the whole expected and attendant predicate into two parts, each with a positive content, one of which is something which the other is not, and consequently only one of which can be the true predicate or determination of the subject.

From this point a course is open to us which at first sight might appear likely to lead to contraries. We might pick out those differents which are most strongly contrasted perceptually. The most general modes of contrast are two, quantity and quality, or degree and kind: but these have each a great variety of modes under them. Take first contrasts of quality. We then obtain a class of differents which are the most strongly contrasted with each other, when referred to a common respect. They would stand as opposite extremes to each other, having between them

CONTRADICTION AND CONTRARIETY. 395

Book II.
Ch. VI.
——
§ 4.
Contraries,
Privatives, and
Relatives.

any number of differents less strongly contrasted, as
for instance, 'Black—Grey—White.' This kind of
contrast, then, is one which admits of degrees; it is
not absolute at the turning-point, the *not*, as in contradictories, 'to be—not to be.' The turning-point
may possibly exist, but owing to the qualitative
character of both opposites, it is difficult to determine
at what precise point of the intermediate percepts it
is to be placed. The extremes alone are perfectly
distinct. The use of contrast in logic is expository,
to explain what anything is, by placing it side by side
with the *positive* things which it is not. And in contrasting any two things, the *general* respect in which
they differ, the genus of which they are differentiations, should always be stated.

But quality and quantity themselves, it may be
said, form a contrast, and modes of the one may be
contraries of modes of the other. True, quality and
quantity are a contrast, but they are not contraries;
and for this reason. Each has a positive differentia
of its own, namely, the material element in consciousness, and the formal element. Feeling is the differentia of quality, Form of quantity. And, so far from
these being mutually exclusive in one and the same
empirical existent, they are mutually requisite, and
each involves the other. There is no existent which
has not both of them. They are ultimate elements
of existents, and stand higher than their determinations as contraries; and they are common to both the
members in every contrariety, as well as in every
other contrast but that between themselves.

Both these ways give us only perceptual contrasts,
not logical contraries, or contradictory determinations.
It is essential to the latter, not only that the exclusion

Book II.
Ch. VI.
——
§ 4.
Contraries,
Privatives, and
Relatives.

between them should be mutual, that is, the affirmation of one involve the negation of the other, or that both should be unable to be *true* together; but it is essential also, that the negation of one should involve the affirmation of the other, or that both should be incapable of being *false* together. And this, notwithstanding that the content of both is positive, or a perceptual determination of an existent. More than contrast goes to contrariety.

This we must seek under the remaining head, that of quantity alone. It will be plain from what has now been said, and since we are driven to percepts to furnish us with the basis of contrariety, that the possibility of such contradictory determinations depends on the possibility of making an *exhaustive* division of any given existent, opposite determinations of which shall be contradictory of each other, or such that the negation of one of them involves the affirmation of the other. For only on the supposition of the existent itself, and of its exhaustive division into two parts, will it follow that, if the existent is not determined in the one way, it must be determined in the other way.

Thus *black* and *white* are not contraries. For to say that a thing is not-black, unless you *mean* by 'not-black' some positive colour, is only another and a confusing way of saying 'It isn't black;' the negative really attaches to the copula. And if we do mean some positive colour by it, then it is only a contrast to 'white,' and its negation does not carry with it the affirmation of 'white,' any more than when it does not signify a positive colour.

The question then is, What makes it possible to divide any given object exhaustively? We shall now

Book II.
Ch. VI.
——
§ 4.
Contraries,
Privatives, and
Relatives.

be at no loss for an answer. It is only in virtue of their formal element, time or time and space together, that objects can be so divided. Time and space alone are capable of an exhaustive division, without residuum. But this division must be applied and enforced by conception, that is to say, the point of division of a percept must be made coincident with the moment of contradiction.

Here it is that we again reap the fruit of that fundamental fact in philosophy, which has been so often insisted on in the previous Chapters, the inseparability of the formal and material elements in all consciousness, and in all empirical objects. The exhaustive divisibility of time and space, the formal element, enables us to conceive empirical objects as exhaustively divided, without leaving any third alternative, any medium or indifferent ground between the two extremes. For we take objects, as existents, with a very high degree of generality; we need not much specialise them in order to be certain of their real existence. There are certain features which all existents must possess, and these features it is which both assure the reality of the particular existent subjects, and enable their exhaustive division, so as to give us their opposite determinations as predicable contraries.

The perception of contraries comes, then, not from any *a priori* furniture of the mind; it comes from experience, but it is an experience universal and uniform, an experience moreover ingrained in us by heredity, just as the fundamental fact is ingrained, from which it springs, I mean the inseparability of the formal and material elements. We find in fact that there are certain alternative distinctions which

Book II.
Ch. VI.

§ 4.
Contraries,
Privatives, and
Relatives.

all men draw, and use unhesitatingly in reasoning, as incapable of a third alternative. These are all cases or modifications of that one fundamental fact; and these distinctions are drawn and applied in the case of empirical objects, because those objects are subject to the laws of the formal element which they contain. Thus it is that those alternative distinctions, which we call contraries, have the same source and the same reason as those primary empirical truths which are the axioms of mathematic pure and applied. But for the present we have not to do with these; it is not in their axiomatic shape, but in their shape as contraries, that we have to do with the primary consequences of the inseparability of form and matter.

The divisibility without residuum of time and space, enforced by the conceptual moment of contradiction, is the source of contrariety. And this is shown by what has been already remarked, that qualities or feelings *per se*, or in their character as feeling, escape classification as contraries; it is only as involved with the formal element that they fall under it. They form perceptual *contrasts*; and these contrasts, when ranged in series, and imagined as carried to a common ideal limit, are sometimes named as contraries. Pleasure and Pain; Black and White; Sad and Merry; Good and Evil; are instances. But these alternatives have an artificial character; we have to make some scheme or scale, some tree or fan or spectrum or sphere, to range them in; and it is also requisite that the series to which they belong should be imagined as ideally complete, and also that the middle state of indifference should be reduced to a point, before the extremes can be taken as repre-

CONTRADICTION AND CONTRARIETY. 399

BOOK II.
CH. VI.

§ 4.
Contraries,
Privatives, and
Relatives.

sentatives of contrary alternatives, one of which must exist if the other does not.

Again, there is no contrary opposition possible between the special senses, or other kinds of feeling, *e.g.*, between sight and hearing. We can neither bring them into a single series, nor imagine intermediates between them which should necessarily belong either to the one or to the other. They are contrasted as modes of feeling; that is the general respect in which they are at once opposed and united. But the whole sphere or group to which these modes of feeling belong cannot in respect of them be exhaustively divided. It is a fiction of the imagination. The same may be said of the contrast between quality and quantity, or of that, still more ultimate, between matter and form, according to what has been remarked above.

Four sorts of apparent contraries, then, must be put aside as mere cases of contrast. First, the contrast between form and matter, or quantity and quality. Second, those contrasts where a series of intermediate feelings between two extremes, of the same kind, can be imagined to exist, without any of them being indifferent, as in *pleasure* and *pain*. Thirdly, those where a group of different kinds of feeling can be imagined, but cannot be completed in imagination, nor rendered exhaustively divisible, as in the case of the special senses. And fourthly, those which have as their point of opposition only the moment of contradiction itself, the *not-this* in its various shapes.

These last will be found to be mere negations of some special modes of feeling, and as such are cases under the general one of Conscious and Unconscious.

Book II.
Ch. VI.
——
§ 4.
Contraries,
Privatives, and
Relatives.

They have no more right to be called contraries than 'black' and 'not-black' have: notwithstanding that the names of many have a positive appearance, owing to the specialness of their existents; as for instance: Visible and Invisible, and their obverse aspect, Light and Darkness; Sound and Silence; Tangible and Intangible; Having Sight and Blind. It is not always easy to discriminate this class from true contraries, when they are very much specialised. *Blindness* for instance is a name for the mere absence of sight in an organism which one would expect to find capable of it. But *Death* is a name for some positive modes of motion in bodies, opposed to other positive modes which are called *Life*. The criterion is, whether *both* sides have a positive content, besides being contradictory.

There remain only, as true contraries, those which are applicable to all existents in their most elementary shape, and are derived from the exhaustive divisibility of Time and Space, the formal element, alone. These may be classed under three heads.

I. Those which are drawn from the division of Time, and therefore apply to all existents without exception, whatever their material elements may be. Among these are:

Former	Latter.
Simultaneous	Sequential.
Static	Dynamic.
Continuous	Discrete.
One	Many.
All	Some.
Even	Odd.
Changing	Unchanging.

BOOK II.
CH. VI.
—
§ 4.
Contraries,
Privatives, and
Relatives.

Determinate	Indeterminate.
Separable	Inseparable.
Same	Different.
Finite	Infinite.

II. Those which are drawn from the division of Space, and apply to existents in space as well as time. Among these are:

Extended	Unextended.
Straight	Curved.
Equal	Unequal.
Whole	Part.
Up	Down.
Right	Left.
Forwards	Backwards.
Uniform	Multiform.
Inner	Outer.
Rest . . .	Motion.

III. Those which are drawn from the division of tangible and visible objects in space and time. Among these are:

Composition	Decomposition.
Active	Inactive.
Plenum	Vacuum.
Cause	Effect.

Such are the classes into which logical contraries seem most properly to fall, together with the most general pairs of contraries under each. They are all positive determinations of existents, and yet contradictory of each other, being drawn from differences in their formal element. Those of the third head are differences which, roughly speaking, belong to Sir

BOOK II.
CH. VI.
—
§ 4.
Contraries,
Privatives, and
Relatives.

W. Hamilton's *secundo-primary*,[1] which are Mr. H. Spencer's *statico-dynamical*,[2] properties of body. And the *primary* and *static* properties of those writers fall, still speaking roughly, under my second head; their *secondary* and *dynamic* under my first head. But no exact correspondence is to be looked for between these two classifications and the present one, founded as they are on different considerations.

The properties of body which give it a place under my third head, and make it the subject of the third group of contraries, are compounds of some of the feelings which are grouped under the second head, and these feelings do not lose their contrary determinations on entering into combination. Neither do feelings in space lose the contrariety which they have as feelings in time. The mechanical properties of body are subject to all the distinctions of its mathematical properties, with others of their own besides. The order in which I have placed these groups is their order in *analysis*, in following the increasing complexity of their formal element.

But when we take these three groups in order of *genesis*, the properties belonging to the first and second are consequences of those belonging to the third. Some mechanical and molecular structure and motion are the conditions of our having any sensation or feeling at all. We then just reverse the order of analysis; mechanical objects stand first, then those called *primary* as belonging to body *per se*, and lastly those called *secondary* as not themselves involving the perception of space, but only resulting from it.

---

[1] Hamilton's Reid, Note D. Vol. II. p. 825.

[2] Principles of Psychology, Part VI. Chap. XI. Vol. II. p. 136. 2nd edit.

Book II.
Ch. VI.

§ 4.
Contraries,
Privatives, and
Relatives.

In either way of taking these groups, we possess in them a framework large enough to embrace all phenomena whatever, any feeling however complex or however simple. There are no phenomena which escape the grasp of the contrary distinctions grouped in this way.

But in the next place, what precisely is the way by which we come to the perception of these contraries? It is this. We bring before us in perceptual redintegration objects, say, in time; and by mental inspection we see that the train of differents is divided by any given point, exhaustively, into *former* and *latter* portions, and that it has no other determinations *ejusdem generis*. So in the case of existents in space; we see in representation that a line in space, if it deviates from a right line, must be a curve. A "broken" line is not one line but many; an angle requires two lines to form it.

Do I then call these contraries intuitive truths, or eternal verities? Nothing of the kind. True, the method of contrariety with its perception of contraries is a necessary process, standing on the same ground as contradiction itself. But it does not follow that all the contraries which we perceive by this means are necessary also. We may be deceived in our representation. True, some of these contraries are the most certain facts known to us, and these are the foundations of mathematic and the sciences. But all have not the same degree of certainty. It may turn out that the content of some of them is apparent only, a fiction of the imagination; which it may well be, without ceasing to be a positive or perceptual content. There may be contrariety between fictions, just as much as between truths.

Book II.
Ch. VI.
—
§ 1.
Contraries,
Privatives, and
Relatives.

Contraries draw their content from perception,
and we may possibly discover by reasoning that some
characteristics, which we perceive and name in the
gross, or as they appear, owe that apparent character
to a want either of delicacy or of vigour in our
senses; whereby a truth of fact as known by reason
is torn apart from a truth of fact as immediately perceived. The content of some contraries may thus be
destroyed as truth, though it still remains as appearance. This happens in those contraries where the
material element enters into the determinations, as
well as into the division between them. For there
we can *think* farther than we can *see;* we can make
divisions in thought which we cannot follow in sense.
Compare, for instance, *Changing and Unchanging*
with *Continuous and Discrete.* Change means change
in *feeling;* the material element enters into the determinations of the pair of contraries. Accordingly we
may discover, that *Unchanging* feeling is an appearance due to the inadequacy of our senses. We can
represent matter changing, where we cannot present
it as changing. *Continuity* and *Discreteness* on the
other hand are determinations of the formal element
alone; and these are in thought as inevitable as they
are in presentation.

To take other instances, *Rest* and *Inaction* may
prove to be fictions, without on that account ceasing
to be positive members of a contrariety. In that
case, *i.e.*, if we have reason to suppose that nothing is
ever at rest, or nothing ever inactive, the contrariety
will still remain as a record of past thought, and a
means of making present thought clear. For the *appearance* of rest and inaction would not be done away
with, by our knowing that it was an appearance only.

Book II.
Ch. VI.
———
§ 4.
Contraries,
Privatives, and
Relatives.

Lastly let it be observed, in this connection, how closely, though unavowedly, contraries are involved in fundamental propositions of science; for instance, in Newton's First Law of Motion, which it would seem impossible to state without assuming them: " *Lex I. Corpus omne perseverare in statu suo quiescendi vel movendi uniformiter in directum, nisi quatenus illud a viribus impressis cogitur statum suum mutare.*" The Contraries are the very machinery by which, the lines upon which, concrete thought moves, in framing any scientific theory or system. They seem to give its direction to thought itself.

The Categories of Genus, Differentia, and Species, are formed in creating concepts out of percepts; Contraries are categories formed in *constructing* concepts. It is a continuation of the same process as in forming genus and species, but the process modified. We have in fact first made three concrete genera, subordinate *inter se*; Objects in time; Objects in space; Visible and Tangible objects. And the several pairs of contraries are sub-genera under each. The pair itself, not either member of it alone, is the sub-genus. The distinction which it rests upon, the *respect* in which its members are contrary to each other, is inseparable from the two members themselves, and together they characterise the phenomena grouped by them; are in fact a genus. There are no higher genera between these and the three to which they respectively belong. For there is no *respect* in which the contraries are contrasted, other than that which receives its name from them. The point of contradiction alone divides them. Language indeed is flexible, and we may speak of *one* and *many* being contrary in respect of *number*, or of number

Book II.
Ch. VI.
——
§ 4.
Contraries,
Privatives, and
Relatives.

being difference of time in respect of one and many, and not (for instance) in respect of former and latter. Again we may speak of *up* and *down* being contrary in respect of *direction*, and equally well of directions being contrary in respect of up and down, direction *meaning* only one or more of these distinctions. In no case is there any respect extraneous to the contrariety itself, which serves it as basis. It is an ultimate phenomenon in thought, immediately arising on the application of the moment of contradiction, not to percepts simply, but to percepts as predicable in judgments.

I pass now to another branch. The determination of the respect in which a term is predicated, of its *secundum quid*, lies at the basis of all logical contrariety. The two other requisites of contraries, the *de eodem* and *eodem tempore*, furnish the next and final division of them. For the whole list of contraries, when each pair is referred to the same subject and at the same time, is also a list of *Privatives*, determinations the presence of either of which involves the absence of the other, and the absence of either the other's presence. In fact, *Privatives* is the name for Contraries, when it is wished to indicate they are contradictory determinations of the same thing at the same time.

On the other hand, the whole list of contraries, when the ultimate respect in which each pair is contrasted is held fast, but the members of it are referred either to different subjects, or to the same subject at different times, the whole list becomes a list of *Relatives*; for still it is true that both the contraries in a pair have a positive perceptual existence. Those terms are relatives which would be privatives if re-

Book II.
Ch. VI.

§ 4.
Contraries,
Privatives, and
Relatives.

ferred to the same subject at the same time. Their contrariety renders them intelligible, their removal to different subjects, or to different times, renders them capable of being truly predicated, both at once. Without contrariety they would be unintelligible; to understand a thing at all you must distinguish it from what it is not; to understand it thoroughly you must follow it up into contrariety; and the knowledge of the one member of a pair of contraries is a necessary ingredient of our knowledge of the other member. The exact sciences are those which are built upon contrariety; and the sciences not commonly called so come near to deserve the name, just in proportion as their phenomena can be exhibited as cases of ultimate contraries; or, in other words, can be brought under the laws of pure and applied mathematic. Throwing contrasts into schemes, or diagrams, is an attempt to bring them as near as possible to contrariety.

As determinations, relatives are mutually necessary. They may be defined as being those contraries each of which, necessarily not being its contrary, exists only when and so far as its contrary exists. Or, in other words, relatives are contraries considered as depending upon each other for their definition, and upon each other's *subjects* as their condition of existing. But it is necessary to repeat the remark which was made about contraries, namely, that there may be apparent as well as true relatives. One of the determinations may be found imaginary, and the phenomena once included in it transferred to the other. The relation then remains only as the record of a mistake; a fate which, it may be observed, is by no means peculiar to contraries and relatives.

Book II.
Ch. VI.
———
§ 4.
Contraries,
Privatives, and
Relatives.

Such is the relation between the three categories of contrariety,—contraries, privatives, and relatives. A familiar instance may serve to exemplify it.

Master and *Servant* are contraries, not indeed ultimate, but still contraries, in respect of a common set of actions, in which the two persons are interested. One commands, the other performs, the same thing. A man cannot be master and servant to the same person at the same time and in the same respect; master and servant are therefore privative terms. They are also relative terms, because you cannot have a master without being a servant, nor have a servant without being a master. Whole and part; cause and effect; half and double; antecedent and consequent; and in fact all the list of contraries; are privatives and relatives also. Whether they are privatives or relatives depends on how they are used. As privative terms they are mutually destructive, as relative terms they are mutually supporting and mutually requisite.

This analysis lets in a ray of light upon Spinoza's *Causa Sui*, Aristotle's αὐτῷ αἴτιον. It is a contradiction in terms. As united in the same subject, cause and effect are *privatives*. Only as separated are they *relatives*. The *Causa Sui* is only the logical hyperbole in which finite thought involves itself in trying to think the infinite object. Spinoza chose to make this logical hyperbole the basis of his system of philosophy. It is surely better to discard such an hyperbole, and simply state the object *to be infinite*, without involving oneself in contradiction by endeavouring to state it in terms of finite thought. We then get simply an Eternal Being, a Being *endless as Time*, instead of a Cause-Effect Being, which is a contradiction.

CONTRADICTION AND CONTRARIETY.   409

Book II.
Ch. VI.
———
§ 1.
Contraries,
Privatives, and
Relatives.

All scientific, as well as popular, thought moves on the lines of the two categories of privatives and relatives, which together exhaust that of contraries. They are the peculiar categories of science. The list of contraries, supposing it complete, would contain all the scientific *genera*, from the highest to the lowest; and the determination of what is at any time the lowest known genus, in any branch of science, to one of its alternatives is the latest particular positive truth at that time discovered. Some, the highest. genera, with their determinations, are discovered already; others await discovery; and the former are the basis and guide to the latter. Science does not come into the field of discovery equipped with all, but only with some few of the entire list of contraries, and those are not separable, but involved in already discovered facts. With these she attempts to discover new facts, which are the determinations of new contraries. And there is one mark of contrariety in every discovered and verified fact, namely this, that it could not have been otherwise.

It is not, then, until we come to the two last categories of logic, privatives and relatives, that we come to modes or determinations of thought which are actually in use in scientific processes. Is this involved in that; if not, what is; what other things does this thing involve, or exclude, by its presence, and what by its absence; how comes, *i.e.*, what are the necessary conditions of, this or that actual experience; what may we expect to find in consequence of finding this circumstance, and what in consequence of not finding it;—such are the sort of questions which I imagine scientific men are constantly putting to themselves, and using to guide experiment. They

Book II.
Ch. VI.

§ 4.
Contraries,
Privatives, and
Relatives.

are engaged, in other words, in more closely, minutely, and fully, determining privatives and relatives. For these categories are not classes made up and done with; they are fixed in their own nature, but indefinitely flexible in their applicability and their content. The more specialised an existent is, the more specialised also will be the contrary alternatives which may be, and one of which must be, imagined as its determinations. And the first step towards answering any of the above questions concerning it is to imagine alternatives affecting its subject.

There is a lacuna, as it seems to me, in this part of Logic, as it is usually treated. It is not made clear what precisely is the mental operation of making scientific discoveries;—making, not demonstrating when made; framing hypotheses, not testing them. For Aristotle's syllogistic, or the several substitutes that have been proposed for it, may stand as a sufficient account of the latter. We are told, indeed, that the process of discovery is tentative, and by means of hypotheses; but the question is, what guides the *tentations*, what is it to frame an hypothesis?

It is now evident that it is an exercise of productive scientific imagination, to which it will be remembered that, in the foregoing Chapter, I promised to recur.[1] It is a process of picturing not of thinking; of construction not of analysis. To picture anything in time and space is to picture it subject to the Contraries; that is, not as chaotic, but as a part of a completed picture of antecedents, consequents, and coexistents, and as changing in correspondence with their changes. The Contraries are time and space relations, which are the formal laws

[1] p. 305.

BOOK II.
CH. VI.

§ 4.
Contraries,
Privatives, and
Relatives.

proper to percepts, just as genus, differentia, and species, are the formal laws proper to concepts.

I have already appealed to the testimony of Professor Tyndall on this point. But it is not only in physical science that the constructive imagination is employed. It is employed equally in mathematic. The facts of the case are, says Professor Sylvester, "that mathematical analysis is constantly invoking the aid of new principles, new ideas, and new methods, not capable of being defined by any form of words, but springing direct from the inherent powers and activity of the human mind, and from continually renewed introspection of that inner world of thought of which the phenomena are as varied and require as close attention to discern as those of the outer physical world (to which the inner one in each individual man may, I think, be conceived to stand in somewhat the same general relation of correspondence as a shadow to the object from which it is projected, or as the hollow palm of one hand to the closed fist which it grasps of the other), that it is unceasingly calling forth the faculties of observation and comparison, that one of its principal weapons is induction, that it has frequent recourse to experimental trial and verification, and that it affords a boundless scope for the exercise of the highest efforts of imagination and invention."

This passage, which occurs in an Inaugural Address to the Mathematical and Physical Section of the British Association, is accompanied, in the edition from which I quote,[1] by a note which gives a detailed account of the steps which led to discovery in a particular instance; among which the employment of

[1] In a volume entitled *The Laws of Verse.* Longmans, 1870. p. 108.

Book II. Ch. VI.

§ 4. Contraries. Privatives, and Relatives.

constructive imagination is conspicuous. "To find the true representation was like looking out into universal space for a planet desiderated according to Bode's or any other empirical law. I found my *desideratum* as follows: I invented a catena of morphological processes which"—But I must refer my reader to the description itself. Here, then, we have the scientific imagination at work; here is a case of Kant's *construction of concepts*.

Such is the point where, and the process by which, general logical thinking becomes scientific thinking, yet without ceasing to be logical. The principle of contradiction, which is the sum and substance of purely formal logic, enters into the generals, particulars, and singulars, into which it has already transformed percepts, and, again coöperating with laws which it does not make but find, transforms them into contraries, privatives, and relatives. This gives us also the true sense of the distinction between pure (or purely formal) logic, general logic, and applied (or scientific) logic. The principle of contradiction in forming genera and species gives general logic; in forming contraries it gives scientific. We have now analysed both operations, both combinations. The mode of operation is essentially the same; in both cases a difference is *thought*, in consequence of having been *perceived*. It was shown that thinking a difference was involved in framing genera and species, classes of similars distinct from their dissimilars.[1] And it is but a further carrying out of the same process, when, supposing such classes formed, we either contrast the extremes of each class by their difference of feeling, or else, taking a difference of

[1] Chapter V, p. 297.

Book II.
Ch. VI.

§ 4.
Contraries,
Privatives, and
Relatives.

form as a basis, oppose its members as true contraries to each other. The difference between the two cases is the following. In forming classes of similars, the *differentia* which distinguishes them is a feature or set of features belonging in each case to the single class which it distinguishes, but throws no light on what the differentia of any other species or class may be. In contraries, on the other hand, the differentia of one species gives a knowledge of the differentia of the other and contrary species under the genus; only two such differentiæ and consequently only two such species of the common genus are possible. This is the meaning of the old rule *Contrariorum eadem est scientia.* We may call the contrast between ordinary logical species *differential*, and that between contrary species *polar.*

I subjoin a Table of Opposition.

The Principle of Contradiction gives rise to Contradictory Propositions, and these to

Contradictory Terms.

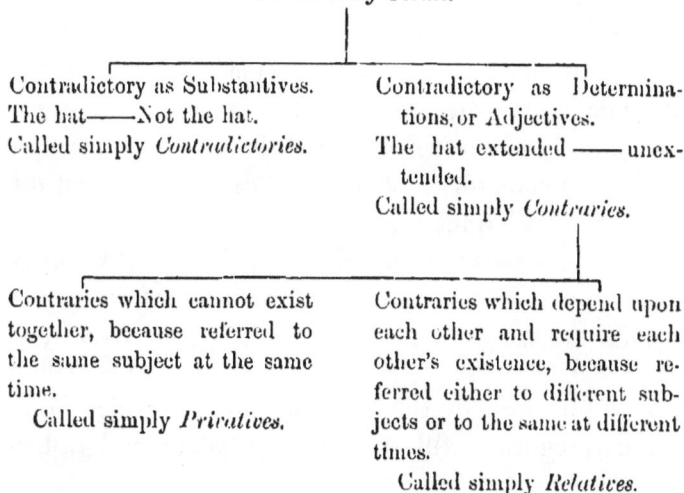

Contradictory as Substantives.
The hat——Not the hat.
Called simply *Contradictories.*

Contradictory as Determinations, or Adjectives.
The hat extended —— unextended.
Called simply *Contraries.*

Contraries which cannot exist together, because referred to the same subject at the same time.
Called simply *Privatives.*

Contraries which depend upon each other and require each other's existence, because referred either to different subjects or to the same at different times.
Called simply *Relatives.*

Book II.
Ch. VI.
—
§ 5.
Final determi-
nation of the
Modal Cate-
gories.

§ 5. When we quit the simple categories of thought for the modal, we step from the domain of science to that of philosophy. The modal categories are those of reflective thought, in contrast with the simple categories which belong to simple, that is, to primary and direct thought. This transition we have to make here, just as we made it in the former part of this whole subject, in the foregoing Chapter. And just as we have found that the principle of contradiction works a modification in the simple logical categories of genus, differentia, and species, by arranging their phenomena under other simple categories, namely, those of contradictories, contraries, privatives, and relatives, so we shall now find that the same principle of contradiction, applied to modals, will modify their phenomena in an analogous manner, and will present us with a new table of modals as the ultimate categories of philosophical thought. Only with this difference, that whereas the categories of Contrariety are not substituted for but added to those of genus, differentia, and species, the modal categories on the other hand, not having been yet finally determined, will be replaced by a new list of modals, under the action of the principle of contradiction. It will be remembered that the consideration of the contradictories of the modals was reserved for the present Chapter.[1]

The modals were found in the foregoing Chapter to be these three, each having a double aspect:

(The Actual.	(The Possible.	(The Necessary.
(The Existent.	(The Contingent.	(The Universal.

The application of the principle of contradiction to these categories will disclose a fundamental differ-

[1] Above, Chapter V. p. 368.

Book II.
Ch. VI.

§ 5.
Final determi-
nation of the
Modal Cate-
gories.

ence between them. Let us suppose each of them to have a contradictory. Thus:

The Actual or Existent.	contradictory	The Non-Actual or Non-Existent.
The Possible or Contingent.	contradictory	The Not-possible to be or not-to-be.
The Necessary or Universal.	contradictory	The Not-Necessary or Not-always-existent.

Dropping, for brevity, the double nomenclature, let us consider the three negative contradictories in this list, the non-actual, the not-possible to be or not to be, and the not-necessary. The non-actual, by which is meant anything that is a representation only, and not a presentation, may clearly, *prima facie*, be either possible or necessary; that is, it may belong to either of the two remaining positive categories. So also the not-necessary may be either actual or possible. But what of the not-possible either to be or not to be? It would seem to be totally and for ever excluded from both the actual and the necessary. I mean that the *content* of the category contradictory of the possible (or contingent) would be excluded from existence altogether; which is not the case with the content of the other two contradictory categories. Their content finds a place in some positive category, either as actual or possible, and either as possible or necessary. A further examination of this singular category is plainly required.

Now observe in the first place, that the alternative, (the affirmative and negative of both the supposed contradictories, both the possible to be or not to be, and the not-possible to be or not to be), the alternative in each is a choice between contradictories, to be—not to be, to be actually perceived—not to be

Book II.
Ch. VI.
—
§ 5.
Final determi-
nation of the
Modal Cate-
gories.

actually perceived. Now take first the contradictory
of the Possible to be or not to be, viz.: the Not-
Possible to be or not to be. If the negative alternative of this contradictory is taken, viz.: Not-possible
not to be, then the contradictory of the Possible becomes a case of necessity, Necessary to be; for the
possibility of not being is excluded. If the affirmative alternative is taken, viz.: Not possible to be, then
it becomes a case of impossibility, for the possibility
of being is excluded. The principle of contradiction
enters, as it were, into the kingdom of the Not-
possible to be or not to be, and divides it between
two contrary alternatives, the necessary and the
impossible.

Take next the positive member of the contradiction, the Possible to be or not to be, itself. The
same breaking up takes place. Take the affirmative
alternative; if a thing is possible to be only (and not
also not to be), then it is a case of necessity, it is no
longer possible not to be; it is *in via* towards actual
existence; it is what is called *potentially* existing,
δυνάμει but not ἐνεργείᾳ. Take the negative alternative; if a thing is possible only not to be (and not
also to be), then it is a case of impossibility, it is no
longer possible to be. The Possible also is, therefore,
a category composed of contrary alternatives.

Now we have seen that contraries cannot exist
together. There is therefore *nothing* to which the
category of Possible (or Contingent) applies. No
content of it exists; and, even supposing it to exist,
it belongs not to the Possible but to the Necessary.
But then, on the other hand, the Possible to be or not
to be, though it is composed of contrary alternatives,
and therefore cannot be regarded as a substantive

Book II.
Ch. VI.
———
§ 5.
Final determi-
nation of the
Modal Cate-
gories.

existent. in the same sense as the actual and the necessary can, yet is not excluded from reflective thought altogether on that account. Experience tells us that it is a category of which we make daily and hourly use. We have seen in the foregoing Chapter that it is a modal reflective category. Let us take the Aristotelian method with it and ask τί σημαίνει ; what do we *mean* when we call anything possible? We mean that, on the one hand we do not *know* of any evidence or *argumentum* against our supposing it to happen or have happened, or of any efficient cause, in order of existence, which should hinder or have hindered it from happening ; and that, on the other hand, we do not *know* of any proof in favour of its happening or having happened, or of any efficient cause which must bring or have brought it to pass.

The Possible to be or not to be, therefore, expresses the relation of a given representation to the present state of our knowledge relevant to it; not, like the actual and the necessary, the relation of objects to consciousness simply. And it consists of a pair of determinations of that relation, or that knowledge, which in their extremes are *contraries*, but shade off into many intermediate degrees of probability and improbability. The field of uncertainty which lies between what is necessary or conditionally necessary, on the one side, and what is impossible or condition- ally impossible on the other, is the field of the Un- certain, or Possible to be or not to be ; an uncertainty attaching to particular states of knowledge, as such, and determined in the direction of being, or in that of not-being, according as our knowledge of the circum- stances is greater or less, but always an uncertainty, an open alternative, until at the limit, either way, the

Book II.
Ch. VI.
——
§ 5.
Final determi-
nation of the
Modal Cate-
gories

possible ceases and either the necessary or the impossible begins.

Now let us note well what it is that has taken place with the category of the Possible to be or not to be, under the action of the principle of contradiction. It is this. We have discovered that it is no determination of existence, no category with a double aspect, subjective and objective, but a determination of our partial knowledge concerning the Necessary. Our partial knowledge is the substantive existent of which the category of the Possible is the determination;—our partial knowledge;—that is, a portion of the subjective aspect of existence, taken apart from the objective, and then treated as if it was an object itself. It is a case in which, as we now discover, we have elevated an inability of consciousness to the rank of a positive attribute of existence, have taken our ignorance of facts for a "contingency" in the facts themselves. For the alternative, to be *or* not to be, is introduced by the partial character, the imperfection of our knowledge at any particular time. We know that whatever 'possible' has been, or shall be, *existent*, is necessary. Assuming the alternative 'to be' concerning it, it *must* either be in the future or have been in the past. This assumption it is which renders its necessity conditional, makes it what is called Potential, instead of simply necessary. While to assume the alternative 'not to be' concerning it, is to make it impossible, and banish it from existence altogether, with all its possible conditions.

The Potential or Conditionally-Necessary, therefore, steps into the place of the Possible or Contingent, under the action of the principle of contradiction on this latter category. Take any particular representation. If it *ever* exist, then it is necessary to be actual sometime, it is necessary *sub modo* or conditionally; the "condition" being the assumption that it exists sometime; its conditions *existendi* will arise, and it will follow. But if it never exist, then it is impossible to be actual at any time; its existence altogether is contradicted. We may call it the conditionally impossible, and its conditions *existendi* will not arise; it is a mere figment or mistake.

Next observe the further light thrown on the two other modal categories by this change in the Possible, and first on that of the Necessary or Universal. A second sense of the term necessary must now be distinguished. Besides the sense in which it is equivalent subjectively to *universal*, or everywhere and always present to consciousness, there is now the sense in which it means *inevitable connection* between partial or particular objects, the *nexus* of thought and the *nexus* of phenomena, the bond which holds together a conditionate and its conditions, an effect and its cause. And here, not in any ready made and *a priori* form of thought, is the origin of the notion of Causality. The inevitable connection between cause and effect is the connection between the parts into which a whole has been broken up; the whole being the Necessary in the sense of the Universal, and the parts being the particular objects which we happen at any time to know, and which, in that character, we call the Potential or Conditionally-Necessary. By reflection we know that all things, taken generally, whatever they may be, are necessary; but of particular things that are not actual, we do not know that *they* are necessary, because we do not know whether we have rightly

Book II.
Ch. VI.
———
§ 5.
Final determi-
nation of the
Modal Cate-
gories.

defined them, whether the particular thing in our thoughts now is a thing consistent with the rest, or not: we do not know, to use a vulgar proverb, whether we have got the right sow by the ear. All we know is, that, *if we have*, it exists necessarily. It is the partial and particular nature of our knowledge which breaks up the Necessary into the Conditionally-Necessary, yet without destroying its necessity; and this breaking up it is, which places the sense of necessary or inevitable connection side by side with the sense of universal, from which it springs.

Next as to the Actual. The new light thrown on this category consists merely in the new relations which are made to appear between it and the two other categories. Considered in the light of the notion of inevitable connection, the Actual is *a case* of the conditionally-necessary, just as this latter is a case of the necessary. That thing, the conditions of which have happened, is actual, is a presentation. And this actual presentation passes into representation and again becomes a part of the conditionally-necessary. It is as if a single facet of a multitudinous-sided solid revolved into the light, and then onwards into darkness again. The order and connection of the facets are the conditionally-necessary, and the characteristics common to all the facets are the necessary and universal. A thing may be necessary or (objectively) universal, as well as actual, if it is an element or aspect pervading all consciousness or all existence. The necessary and the actual do not exclude each other. And a thing is conditionally-necessary if it is a particular something which has had or will have an actual, presentative, existence some time or other. It is actual when, and while, it exists presentatively. And this is the moment which we seize, the moment of our conscious connection with the universe, the basis and the test of all our thoughts. The necessary itself is *actually* necessary, for not only is the moment of contradiction an actual moment, but Time, the general element of all consciousness does not cease to be present *now* because it is present always.

The character of the three Modal Categories, as we have now determined them, namely:

> The Actual,
> The Conditionally-Necessary,
> The Necessary,

is to be perceptual aspects of existence, or of consciousness, not exclusive of one another, but still aspects which, as such, are partial. They are percepts, terms in their first intention, which nevertheless have passed through the crucible of thought, and have been formed by means of terms of second intention. They characterise existence as it is known to us, and they each characterise the whole of it. They are the specially philosophical categories; while the simple categories of thought, *genus, differentia,* and *species,* with their derivatives, the categories of contrariety, are common to science as well. All these simple categories are applicable only to the second of the modal categories, the conditionally-necessary. It is the task of science to discover two things concerning particular things which are not actual, 1st, *What* particular things exist; 2nd, *Whether* this or that particular thing exists; or in other words, (1) to frame true definitions of particular things, and (2) to discover whether the things defined by it exist. But

Book II.
Ch. VI.

§ 5.
Final determi-
nation of the
Modal Cate-
gories.

science does not investigate either the Actual or the Necessary. These it assumes, and devotes itself to the partial, or Conditionally-Necessary.

§ 6. Such is the way in which the principle of contradiction operates upon both the simple and the modal categories. But it remains to be seen more precisely the mode in which this is effected. This, it will be remembered, was postponed till after the exposition of the modal categories.[1] There are in fact certain categories of a purely logical kind, which we have not yet examined, and which have always introduced great confusion into this whole subject, in consequence of neglecting the distinction between the perceptual and conceptual orders. I speak of the categories of the *logically* possible and impossible, the possible and impossible to *thought*, as distinguished from the possible (and impossible) to be or not to be.

The question is, What is meant by the possible to thought, as distinguished from the possible to be or not to be? We have seen that there is no such thing as the possible not to be; and that the possible to be or not to be is, in truth, the potential or conditionally-necessary. But the possible to thought,—what of that? Is that also conditionally-necessary? Everything that we can construe to thought,—necessary? No one can suppose it.

Again, the possible to thought appears to have its contradictory in the *impossible* to thought; and the impossible is defined as a thought which involves a

[1] Above, at p. 392.

CONTRADICTION AND CONTRARIETY. 423

Book II.
Ch. VI.

§ 6.
The Logic of
Possibility.

contradiction. But the necessary is defined as a thought, the contradictory of which involves a contradiction; or, in other words, as the contradictory of the impossible. Accordingly, the impossible appears to have *two* contradictories, the possible and the necessary, which is absurd.

The solution of these puzzles, and the establishment of a short and simple Logic of Possibility, will be effected by considering the moment of contradiction. This moment may be said to be the second centre of philosophy, the first being the moment of presentative perception. They may be compared to the foci of an ellipse; and what depends on the one must not be confused with what depends on the other.

The moment of contradiction, we have seen, is that upon which all thought, all conception, as distinguished from perceptual processes, hinges; and this moment it is which is the moment of necessity, or necessary connection in thought. But now I remark, that it is this same moment which is the moment of *possibility* as well. For, being the moment of contradiction, it shows that one of two alternatives must be true and the other false, but it says nothing whatever to determine *which* of the two is the true, and which the false alternative. This determination comes from another source altogether, from perception, which gives the content of both members of the alternation. The moment of contradiction itself is empty of all content, being merely the point of time determined by attention. And consequently you cannot have the moment of contradiction in consciousness, without assuming, *for that moment*, that either alternative may be true, as well as that one must be true and the other false. This is what is meant by logical possibility. You cannot have contradictory alternatives posed, without the possibility of either alternative being assumed, to begin with, and so far as the logical principle of contradiction goes. Otherwise, reasoning would be a process involving foregone conclusions.

The possible to thought or logically possible, then, may be defined *provisionally* as a concept which neither itself, nor the contradictory of which, involves a contradiction; but then it must be added that this 'logically possible' exists only while we make the proposition which expresses it, or as long as our state of complete uncertainty exists. The logically possible has no existence other than as the object, not of our certainty, but of our uncertainty; the undetermined state of our minds is objectivated in it; and whenever a concrete concept, or concept with a content, is called logically possible, it can only be when it contains no mark whatever to show whether it would be truly or falsely predicated in a particular case. It would then be a case of the logically possible, because it would be a case of indetermination, a supposed object of the moment of our logical indifference or balancing.

In the next place it is to be observed, that existence, the stream of percepts, is pre-supposed in logical possibility, and does not enter into question. The primary question is not *whether* this or that exists, but whether this or that *is so and so;* for instance, whether the hat is black or not, not whether it exists at all. That something exists in consciousness at the time is pre-supposed, and the question is—*what* exists there. Is it black, is it white, is it *a hat* at all? The circumstance that thought pre-supposes perception is that which determines this restriction of the logically possible. And we have to do, in the first operations of thought about the logically possible, not with the existence or non-existence of substantive things, but with their adjectival determinations, that is, with contraries not with contradictories.

For let any one attempt to put to himself the question of the possibility that *Nothing* exists,—he will find he cannot do it. While he is conscious at all, he is conscious of existence. He will find that he is imagining, not Purum Nihil, but merely a time or space vacuum. Therefore it is that the moment of logical possibility pre-supposes existence of some sort or other; and in starting from this supposition, we aim at determining *what* this existence is.

Here is the place, and here also it seems necessary for obviating objections, to interpose an answer to a singular argument of Mr. Herbert Spencer, where[1] he points out what he thinks "the contradiction between the assertion that consciousness cannot be transcended and the assertion that there exists nothing beyond consciousness." These assertions seem to me identical, two ways of stating the same thing. But Mr. Spencer proceeds: "For if we can in no way be aware of anything beyond consciousness, what can suggest either the affirmation or denial of it? and how can even denial of it be framed in thought? The very proposition that consciousness cannot be transcended, admits of being put together only by representing a limit, and consequently implies some kind of consciousness of something beyond the limit."

[1] Principles of Psychology, Part VII. Ch. 14. Vol. II. p. 441. 2nd edit.

By representing a limit.—yes; but not a limit *to consciousness.* That is the very point in question. Within consciousness arises the method of setting limits, by means of contradiction. It is this method, this use of contradiction, which "suggests affirmation or denial," and enables "denial to be framed in thought." But try to apply this method to consciousness itself as a whole, and you find you cannot do it; try to apply it to existence as a whole, and you find you cannot do it. Now it is this inadequacy of logical limitation to consciousness and to existence which is expressed in the assertions in question. The method of contradiction or limitation arises *within* consciousness and existence, and its use pre-supposes them. Mr. Spencer, I imagine, means to apply the limit to consciousness, and then supposes himself to find existence beyond it; I take him to be an *objective absolutist.* On the other hand, it is equally illogical to apply the limit to existence, and then find consciousness beyond it, which is the way preferred by *subjective absolutists.* But to return.

In the respect now mentioned, the Logic of Possibility does not differ from the positive logic which we have hitherto been examining; both alike pre-suppose perception and existence. But now a remarkable difference between them must be noted. The principle of contradiction is the common centre or starting point of both; but we have seen that this principle has two aspects, involving each other, that of necessity and that of possibility. What I have called the Logic of Possibility is based upon the latter aspect of the principle; by which I mean that we make use of the logical indifference, involved in contradiction, in order to assume, *for the purposes of*

CONTRADICTION AND CONTRARIETY. 427

BOOK II.
CH. VI.

§ 6.
The Logic of
Possibility.

argument, either member of a contradiction to be true.

But this liberty is practically exercised only by assuming to be true that member which we design to disprove. For to assume the truth of what you are about to prove is a *petitio principii*. Unless indeed to this extent, that you use the assumption, not as a premiss, but only as a guide to investigation by further observation or experiment; or else that you show the agreement of what you so assume with otherwise known facts,—a process which is good as corroboration, but cannot extend to demonstration. For the known facts might possibly agree with and be conditioned by other things as well as that with which you will have connected them.

The special principle, then, of the logic of possibility is the postulate of contradiction, just as that used in concrete positive thinking, in framing and using genera and species, is the postulate of identity, the formula of assertion in which is, A is B, as *e.g.*, The hat is black. The logic of possibility on the contrary tests truth, not directly by asking what we find in experience, but indirectly by excluding what we cannot find there. The rock which serves it for leverage is the notion of the *Impossible* to thought, that is, the Self-contradictory. And its method is the method of *apagogic* as distinguished from *ostensive* reasoning, a method by which we can test the validity even of our primary perceptions, namely, by coming back upon them through the assumption of their contradictories, which are seen to lead inevitably to conclusions which are false in fact. This is the gist of apagogic reasoning, or the *reductio ad impossibile*.

The first category, then, which is established in this use of the postulate of contradiction is that of *The Impossible to thought*; usually defined as that which involves a contradiction; but which is truly defined, in accordance with what has now been said, as *that which involves privative determinations*, e.g., *a round square*. Round and Square, as determinations of the same figure at the same time, are privatives.

The next category is the *contrary* of this, *The Necessary to thought*. It is usually defined as that the contradictory of which involves a contradiction; but is more properly now to be defined as *that the contrary of which involves privative determinations*, e.g., *a curved circle*. Why? Because its contrary, an angular circle, is impossible.

The third category is the indetermination of both the former, *The Possible to thought*. Its usual definition has been given, as provisional, above.[1] But its true definition as a supposed object is, *that which neither itself, nor the contrary of which, involves privative determinations, e.g., a round figure; a square figure.*

More briefly, the *Impossible* is whatever cannot be construed to thought, the *Necessary* whatever must be in thought if any part of it is, the *Possible* whatever can be construed to thought, but need not be so, because its contrary can be construed to thought equally well. These are the three categories of the Logic of Possibility, and the basis of Apagogic Reasoning.

Observe that in this scheme the *Necessary* to thought and the *Impossible* to thought are true contraries, and exhaust between them the whole field of the Possible to thought, which sinks to a mere point of division without any content. They are contraries in respect of being and not being, or of existence simply.

But we may also consider the Possible to thought as a concrete object, the balancing point objectivated, and then the Necessary and Impossible to thought are *together* opposed, as a contrary, to the Possible; the respect in which this pair of contraries is made being the respect of determination. It is a case of the more general pair of contraries, the determinate and indeterminate. But this way of taking the matter rests on the fiction of making a supposed object out of the mere point of indetermination. The logic of possibility rests on keeping that point a point merely; and its true contraries are the Necessary and the Impossible to thought.

In fact we have, in the Logic of Possibility, the subjective side of the method of thought by which Contraries are formed. And all the pairs of contraries which were named above, or which can be named and brought under them, are concrete *cases* of the three categories of the logic of possibility, or, what is the same thing, of its pair of logical contraries, the Necessary and the Impossible.

And now this question occurs;—What is the relation of the logically possible to the modal category of the possible to be or not to be? We have arrived at them by quite different routes, we have kept them carefully distinct. But I shall be surprised if any one can see, after all, any difference between them. They seem to me identical. And for this reason. We have shown that, in the logic of possibility, existence is assumed, some content or other is assumed as existing. A round figure may exist, a square figure may exist. The question only remains, whether they can exist here and now, in this or that actual connection. The question about them is no longer *what*, but *whether* they exist: the second of the two questions which can be asked about anything, the question of its *conditions* either *existendi* or *cognoscendi*. *Possible to thought* means, in its positive alternative, construable to thought, possible *per se*; *possible to exist* means, (supposing construability to thought), having determining conditions to bring it about.

Now this is precisely what the Possible to be was shown as; for it was shown to be replaced by the Potential or the Conditionally necessary. We are again on the old ground of the Modal Categories. And the whole Logic of Possibility moves within the limits of the conditionally-necessary, the partial aspect of existence, just as the logic of contraries, privatives, and relatives, in its concrete and ostensive branch, was shown to do.

When we come to the second question relating to anything possible to thought *per se*, namely, the question *whether* and in what connection it exists, we come to the further categories of the Probable and Improbable. The thing *de quo* is then taken as represented by its definition, and as such supposed to be possible. The possible to thought is then treated as the Contingent, as something which may or may not exist, an event which may or may not happen, according as circumstances favour or prevent it, these circumstances being in their turn contingent also. There are thus degrees of contingency, and these degrees are degrees of Probability or Improbability, that is, of more or less certainty or uncertainty

CONTRADICTION AND CONTRARIETY. 431

Book II.
Ch. VI.

§ 6.
The Logic of
Possibility.

in our knowledge. And they are opposed to each other as more and less, that is, as *contrasts* not as contraries. They shade off into one another almost imperceptibly, as greater or less intensities of belief; and apart from the degree of our knowledge about the facts there is no contingency about them, but everything is necessary or conditionally-necessary.

The contingent and its two main contrasts may be thus defined :

The Contingent : Possibles to thought, the conditions of which involve no privative determinations, and the conditions of the contraries of which involve none.

The Probable : Contingents for which some evidence can be shown, or more weighty than can be shown against them.

The Improbable : Contingents against which some evidence can be shown, or more weighty than can be shown for them.

The contingent, with its probabilities and improbabilities, is the particular portion of the logical field which is occupied by what is known as the Theory of Probabilities or Chances. The second of the two questions which arise under logical possibility, namely, not the question *what* things are possible, but the question *whether* a definitely described thing or event occurs or does not occur,— this is the field of that theory. In it, the thing or event *de quo* is always definitely envisaged, or, if it is taken as a thing or event capable of happening in several different ways, still these are definitely envisaged and distinguished from others. No question arises about *what* the event or thing, or its modes of

occurring, are. These being supposed given, what is
aimed at is to determine what are the chances for or
against its occurrence, or its occurrence in a particular way; and according to these the otherwise
indefinite intensities of belief are determined, or
sought to be so. For instance, the chances for and
against A B living ten years more, the chances for
and against ten trumps in a hand at whist.

This is a part of logic in which mathematical
methods are perfectly legitimate. The *nature* of the
things dealt with being supposed known, all that
remains to deal with is their *conditions*. Of these,
again, whatever can be known by observation and
experiment belongs to what may fairly be called the
inductive basis of the doctrine of chances.[1] Whatever is inductively known is no longer thought of
as contingent. There remain for the calculation of
chances itself just the results of those minute or intricate conditions which observation and experiment
cannot reach, as for instance, those mechanical motions which determine the throws of fairly-made dice.
These are no doubt laws of nature, but our knowledge of them is knowledge in the lowest stage of
empiricism, being in the form of statistics, that is,
facts stated without their conditions.

And now one general remark in conclusion of
this branch of the subject. It is, that, in virtue of
the facts upon which is based the Logic of Possibility, philosophy has a certain *negative* voice in the
questions of positive science. For in its application
of this kind of reasoning, it bases itself ultimately
upon certain facts of consciousness which cannot be

[1] See on this point Mr. Venn's excellent treatise, *The Logic of
Chance*, Chap. VIII. p. 184. 2nd edit.

denied or reversed, such for instance as the necessity that colour should be extended. It makes its deductions as to possibility, impossibility, and necessity, from facts of this kind, from cases, I mean, of the inseparability of the formal and material elements, together with the principle of contrariety, which, as we have seen, is that which enables the Construction of Concepts. And the construction of concepts is the basis of all scientific theory. This accordingly is a point at which we see the nature and source of one of the perennial contentions between science and philosophy. For this negative voice has very often been wrongly used, from mistaking for ultimate and indissoluble connections those coexistences or sequences which were due only to association and habit, or to superficial acquaintance with the phenomena.

J. S. Mill was perfectly justified in insisting on the many cases in which the assertion of inconceivability, as an argument against new doctrines, or new facts of science, has been found false. But he is surely going too far when he rejects the principle of the inconceivability of the negative as the ultimate test of truth. All my readers will be familiar with Mr. Herbert Spencer's masterly proof of the opposite doctrine, under the title of the Universal Postulate. Nevertheless Mill's caution was needed, and particularly, as it seems to me, in the case of some of the doctrines which Mr. Spencer makes the Universal Postulate to cover with its ægis. It is one thing to see that a principle is true, and another to distinguish aright the cases which fall under it. And Mill himself seems to concede all that I at least should contend for, in the following passage from a Chapter in one of the later editions of his Logic,[1] a chapter which is chiefly devoted to the controversy with Mr. Spencer. "In the few cases in which uniformity of experience does amount to the strongest possible proof, as with such propositions as these, Two straight lines cannot enclose a space, Every event has a cause, it is not because their negations are inconceivable, which is not always the fact; but because the experience, which has been thus uniform, pervades all nature. It will be shown in the following Book that none of the conclusions either of induction or of deduction can be considered certain, except as far as their truth is shown to be inseparably bound up with truths of this class."

Thought is no faculty ready furnished either with ideas or with categories of its own *a priori* to experience. It takes its whole content from perception, of the course of which it is a modification under the action of attention. The ultimate facts of perception are therefore also the ultimate facts of reasoning. The content and nature of these facts are the inevitable and necessary basis of all trains of imagination and argument. The "negative" which is "inconceivable" is the negative of these facts. It becomes therefore of the last importance to discover by analysis what the content and nature of these facts are, and which are the facts that are strictly and truly ultimate. And this is the only safeguard against that class of mistakes which has sown so much discord between science and philosophy, the mistaking mere effects of association for ultimate facts the negative of which is really, and not apparently only, inconceivable.

---

[1] A System of Logic. 6th edition, 1865. Book II. Ch. VII. Examination of some opinions, &c. Vol. I. p. 300.

§ 7. And now I hasten to the conclusion of the present Chapter. Philosophy has been called the Science of the Possible, *scientia possibilium.* I think I have shown in what sense this is a true title. It has also always been held by its disciples to be a doctrine of the *necessity* of things, of *necessary* exist-ence. The sense in which it really is so has now been made clear; it is in consequence of the function of Reflection in consciousness, and of the modal categories which are its self-made instruments and furniture. I do not see what can escape the sweep of these Modals. I see much which escapes that of the simple categories of thought, for instance, the determination of the term *existence* itself. Yet they depend, as has been shown, upon the principle of contradiction as one of their bases; their objective, or more properly speaking, their complete and exhaustive, validity depends on the complete and exhaustive validity of that principle. Is or is not the principle of contradiction, formulated as we have seen by the postulates of logic, absolutely valid? This is one form of a question which requires another opportunity than the present to enter upon in an ex-haustive manner.

Reserving this question for a future chapter, it must be remarked in conclusion, that we have in the modal categories the full and final explanation of the great philosophical problem—In what does the *nexus* of things consist? or, What is the causal relation between them? Not that this nexus, this causality, comes from the modal categories, but that they show us where it does come from, in showing us how the notion of necessity is formed. It is the concept which gives back, conceptually and reflectively modified, the *continuity of Time*. Simple conception breaks up the continuity of the partial perceptual redintegrations, and classifies the pieces, each continuous within itself though discrete from others. Reflection brings back again the perceptual order, enriched by conceptual knowledge, an order no longer partial but forming a single chain of things and events, an image or train of images figuring the entire course of the known universe. For presentations are known and characterised as *actual* in reflective conception. Their *meaning* in perception is presentative percepts or existents. The second list of modal categories exhibits these actuals as cases of the necessary. And they do not cease to be percepts when they pass into representation. The nexus between them is the same from the first; conception adds no new nexus, though it breaks up the partial series of percepts, and enables a completer series to take its place. The nexus is the nexus of perception. The continuous duration involved in percepts and in chains of percepts is inseparable from them, and, when these are supposed to be separated from one another, appears as the inevitable nexus between them.

It is only on supposing things to be separate that the question of their nexus arises. Once see that things cannot be entirely separate, that time which is continuous is *common* to all, and the question of nexus is answered before it is asked. It is a case of what should be called FERRIER's theorem, from the clear way in which it has been stated, and due emphasis laid upon it by him,[1] a theorem which, to

---

[1] Institutes of Metaphysic. Epistemology Prop. III. Ontology Prop. IX. with the Observations and Explanations. The clearest and most separate statement of the principle is perhaps to be found

match with Ockham's famous razor, and also that it may be seen for itself, independently of that use which Ferrier makes of it, it may perhaps be permissible to turn into Latin: *Frustra quæritur quomodo inter ea conjunctio efficitur quæ non nisi conjuncta reperiuntur.* You can point the nexus out in the things themselves; it is given when they are; you can show in what it *consists*, but are precluded from asking how it *arises*.

The notions of efficiency and force are only requisite, and indeed only admissible, when we conceive phenomena as separately existing things, whether spiritual or material, rounded-off substances like pie-crust patterns cut out by the cook. Then indeed the continuity of thought makes itself felt, in forcing us to ask what *connects* these rounded-off things with one another. We have ourselves destroyed the nexus, which we now seek to restore by the fictions of efficiency and force. Physical, popular, notions idolised, a further enforcement and elaboration of the separatist view of things, entities made out of abstractions, these have long done duty for philosophy; a supposed perfecting of the separatist theory for the institution of the analytic; the methods and divisions of science, founded on direct perception, for the methods and divisions of philosophy, founded on reflection.

The earliest origins of philosophy are illustrated,

---

at the end of the latter passage, p. 510. 2nd edit. 1856: "The contradictory elements" (contradictory when taken separately, *per se*) "are found by an analysis of the synthesis, but the synthesis is not generated by putting together the parts obtained by the analysis, because these parts can be conceived only in relation to each other, or as *already* put together."

and its intuitions justified, by this analysis of the modal categories and their reference to reflection. For instance, let any one read by their light the fragments which remain to us of Parmenides, and he will see at once that the " two ways," which Parmenides there contrasts, the "way of truth," and the " way of opinion," are, the first the method of reflection and philosophy, the second that of direct perception and science. This explains what a distinguished literary critic[1] has lately found inexplicable in Parmenides, his proceeding to give an account of " opinion," τὰ πρὸς δόξαν, immediately after insisting that it is not the "way of truth." For it is an account of what men have made of truth, of the modes in which they have broken it up into rounded-off things, constituted by negation, being negations of everything else, having negation for their essence, and yet proclaimed as substantive existents. This is shown by the way in which Parmenides makes the transition from the ὂν to the μὴ ὄν.[2]

οὐδὲν γὰρ ἢ ἔστιν ἢ ἔσται
ἄλλο παρὲκ τοῦ ἐόντος · ἐπεὶ τόγε μοῖρ᾽ ἐπέδησεν
οὖλον ἀκίνητόν τ᾽ ἔμεναι, τῷ πάντ᾽ ὄνομ᾽ ἐστίν
ὅσσα βροτοὶ κατέθεντο πεποιθότες εἶναι ἀληθῆ
γίγνεσθαί τε καὶ ὄλλυσθαι, εἶναί τε καὶ οὐκί,
καὶ τόπον ἀλλάσσειν, διά τε χρόα φανὸν ἀμείβειν.

On the other hand, it is the unity and necessity, given as we now see by reflection, as opposed to direct perception, that Parmenides proclaims as the attributes of Existence, as the objects of the " way of truth."

---

[1] Mr. J. A. Symonds, The Eleatic Fragments. In Fortnightly Review, August 1873.

[2] Fragmenta Philosophorum Græc. Vol. I. Parmenidis Carm. Rel. v. 96. Didot, Paris.

The poem is full of eloquent passages in which the conviction of this unity and necessity, of the infinity and yet completeness of existence, of its continuity, of the simultaneity and coexistence of its parts, is asserted. Such passages must be read in the light of later philosophical analysis; and only then will they appear to have a meaning. We are not to attribute to them, as their own, the meaning to which they point, and which now throws back its light on them. Parmenides is not a post-Kantian; and yet it is only post-Kantians that can see the true meaning of Parmenides. For his meaning and theirs is the same, in respect that both insist upon the method of reflection in contrast to direct or separatist ways of thought. Why separate, says in effect Parmenides, why seek for a connection between things which, if you saw them aright, are never otherwise than connected? Ferrier's theorem is virtually contained in the lines :[1]

πᾶν δὲ πλέον ἐστὶν ἐόντος ·
τῷ ξυνεχὲς πᾶν ἐστιν, ἐὸν γὰρ ἐόντι πελάζει.

And the doctrine of necessity in the lines which immediately follow:

Αὐτὰρ ἀκίνητον μεγάλων ἐν πείρασι δεσμῶν
ἐστὶν ἄναρχον, ἄπαυστον, ἐπεὶ γένεσις καὶ ὄλεθρος
τῆλε μάλ' ἐπλάγχθησαν, ἀπῶσε δὲ πίστις ἀληθής.

We find, then, in Parmenides the distinction between the necessary, universally existent, permanent, world, and the world of ever changing existences, of birth and death, of appearances and disappearances; the one the world of *truth*, the other of *opinion*. This distinction we now know is drawn by reflection, and corresponds to the distinction between

[1] Fragmenta Philosophorum Græc. Vol. I. Parmenidis Carm. Rel. v. 80.

what falls under its category of the Necessary and
Universal on the one side, and its categories of the
Actual and Conditionally-Necessary on the other side.
But again we find also in Parmenides, mixed up with
this, a doctrine which is untenable, namely, the distinction between the One and the Many, used as the
basis of philosophy; the theory is untenable, in contradistinction from the method. And why? Just
because the distinction, as we now see, is one of
direct and not of reflective thought. This is merely
saying that the Eleatics had not obtained a perfect
grasp of that reflective method which they were, perhaps, the first to institute.

Their two worlds undergo, then, in modern
thought a certain transformation. We now see their
inseparability. We can understand that their world
of transitory appearances, being the actual and conditionally-necessary, is not a world of *opinion*, but is
the world of *science*; while their world of *truth* is the
world which specially belongs to *philosophy*. Nor is
there any essential difference between them. One is
not a phenomenal and the other a noumenal world;
but both are phenomenal. That which to us is Past,
and that which to us is Future, are necessarily existent, and are necessarily thought as existent by
reflection. But they are no *Thing-in-itself* on that
account. Nor does it make the least difference
whether we imagine the conditionally-necessary phenomena coming into existence and leaving it again,
as they first begin and then cease to be actual presentations, or whether we imagine the whole necessary phenomenal universe as existent, and illuminated
by a travelling gleam of consciousness, as by a sun-
beam falling upon successive spots in a landscape,

and exhibiting in presentative perception those objects upon which for a moment it rests.

True, it is not easy to imagine in detail the mode in which such arising into actuality takes place, or the mode of existence of things before and after that arising. These are problems of the Constructive Branch of philosophy. Metaphysic has merely to indicate and place them, to show their origin in the modal categories, and the solution which they apparently anticipate. For so much at least can be said *a priori* concerning them, that the consciousness which should see perfectly how *that* takes place, which is the object of these problems, would be no less infinite than the universe itself, would be a consciousness to which all that is possible and all that is necessary would be known in actual presentation.

END OF BOOK II.

LONDON:
ROBSON AND SONS, PRINTERS, PANCRAS ROAD, N.W.